Palgrave Studies in Media and Environmental Communication

Series Editors
Anders Hansen
School of Media, Communication and Sociology
University of Leicester
Leicester, UK

Steve Depoe
McMicken College of Arts and Sciences
University of Cincinnati
Cincinnati, OH, USA

Drawing on both leading and emerging scholars of environmental communication, the Palgrave Studies in Media and Environmental Communication Series features books on the key roles of media and communication processes in relation to a broad range of global as well as national/local environmental issues, crises and disasters. Characteristic of the cross-disciplinary nature of environmental communication, the books showcase a broad variety of theories, methods and perspectives for the study of media and communication processes regarding the environment. Common to these is the endeavour to describe, analyse, understand and explain the centrality of media and communication processes to public and political action on the environment.

More information about this series at
http://www.palgrave.com/gp/series/14612

Mark Terry

The Geo-Doc

Geomedia, Documentary Film, and Social Change

Mark Terry
Humanities
York University
Toronto, ON, Canada

Palgrave Studies in Media and Environmental Communication
ISBN 978-3-030-32510-7 ISBN 978-3-030-32508-4 (eBook)
https://doi.org/10.1007/978-3-030-32508-4

Cover design: eStudioCalamar

This Palgrave Macmillan imprint is published by the registered company Springer Nature Switzerland AG.
The registered company address is: Gewerbestrasse 11, 6330 Cham, Switzerland

Foreword

Environmental themes have been proliferating in film and media alongside the growth of awareness of the scope and scale of the ecological crisis—indeed, the multiple crises in relations between industrial society and its surrounding and intermeshing worlds. As filmmakers and documentarians wield their cameras, critics and theorists have helped to make sense of their work, interpreting it within the context of a changing, and rapidly digitalizing, global mediascape.

Mark Terry's work on the "geo-doc" contributes to this movement in two important respects. The book presents a detailed and fascinating history of the non-fiction film in its capacities to promote and contribute to social change. Terry's history of the uses of non-fiction film extends from the beginnings of moving-image technology to today's digital era, with its i-docs, web docs, and multilinear and database documentaries. The book's earlier chapters include a particularly helpful overview of the Canadian tradition of non-fiction filmmaking, adding insight into our understanding of the pre-Grierson era, including promotional films used to attract British and Ukrainian settlers to the Canadian West, as well as educational, commercial, and proto-ethnographic films.

Secondly, the book proposes and defines a new form of documentary film, the "geo-doc," illustrating the workings of this form through an account of the author's own collaborative documentary film project, the *Youth Climate Report GIS Project*. Terry lucidly explicates debates over "truth-telling," "authenticity," and the virtues of the participatory mode of documentary, and builds on the latter case to incorporate the novel possibilities of digital and locative, geographically based technologies, resulting, in the final chapter, in a clear articulation of the hybrid form that the author identifies as the "geo-doc." Along the way, he provides an interesting analysis of ecocinema, highlighting the special and innovative place of the latter within the recent history of documentary filmmaking.

This book is a richly detailed study of the changing potentials of documentary filmmaking for contributing to social and policy changes. Its assessment of the technical means available to today's digital documentarian is lucid and helpful. Underpinning these critical contributions is Terry's professional experience as a documentarian working with multiple collaborators and audiences, including with the United Nations Environment Programme and the United Nations Framework Convention on Climate Change. As a result, his theoretical deliberations are grounded within a thorough understanding of the medium and its real-world capabilities. The combination of these two strengths—substantial professional experience and well-argued and well-researched scholarship—makes for a rare and admirable synthesis.

Burlington, VT, USA Adrian Ivakhiv

Preface

The reader should know that the author is not only a scholar of documentary film, but an active practitioner as well. My first documentary was made in 1989 for Paramount Pictures and was called *Clive Barker: The Art of Horror*. I produced a simple film showcasing an interview with the multidisciplinary creator of horror fiction with cutaways to samples of his impressive and unique body of work in film, novels, art, and comics.

This film was a natural extension from my original career as a journalist. When I was in my final year as an undergraduate at Toronto's York University (Glendon College), I worked full-time as a copy editor at the *Toronto Sun*, a daily newspaper. When I graduated, I moved to the *Toronto Star* to work as a reporter. These experiences taught me the importance of getting both sides of every story in the pursuit of fair reporting and balanced journalism. It was not our job to choose a side and cherry-pick the information that supported that bias. That was known as editorializing. Today, we see more and more of that in what has become accepted and conventional journalism, but that's an investigation for another book.

The other thing I learned was the corporate importance of grammar and spelling. There were specific styles to which we had to adhere, and if a mistake ever made it to print, it could cost us our job. Again, typos and informal colloquialisms find their way into today's published material at what I consider to be an alarming rate. I suspect the Internet has accelerated the deadline for print journalism and its digital cousin resulting in missed errors caused by rushed editing. What I find uncomfortable is the tolerance of this among readers, even among trolls who usually rejoice in pointing out the mistakes of others.

I sound the alarm bell here not to sidetrack the reader from the subject of this book, but to inform the reader of my pre-academic careers as a journalist and filmmaker and how these disciplines of practice steered my research in terms of evolving the documentary film, generically and technically, into a new form that demonstrates a proclivity to inform and influence power more efficiently than its predecessors.

The Art of Horror, while released theatrically as a documentary short, followed the production regimen of broadcast journalism featuring the standard components of narration, interview, and B-roll. Thus began my documentary film career. In the late 1990s and early 2000s, the documentary film industry was not encouraging social justice crusades as much as we see in most documentaries today. We can trace this shift in perspective to the undeniable popularity of Michael Moore's brand of filmmaking—bookmarked by the unprecedented box office success of *Fahrenheit 9/11* in 2004. This opened the door for many more documentary filmmakers to a new source of revenue beyond the television broadcast licence fee. Until this moment in film history, a theatrical release was rare, and significant box office revenue even more. During this period, the documentary film industry was focused on two content models: Top 10 lists and biographies, and, to a lesser degree, historical stories. This was the preferred content of such broadcasters as the Discovery Channel, Biography, the History Channel, Discovery Wings, and the Military Channel. Stories within these categories resonated with subscribers, and therefore, it became the preferred content of the industry's broadcasters. If a documentary filmmaker wanted to make a film, they usually responded to what the market would buy; so, many documentarians like myself made biographies (*Clive Barker: The Art of Horror*, 1989), Top 10 lists (*The Natural Wonders of the World*, 1999), and historical stories (*We Stand on Guard*, 2000). I was able to pre-sell each of these titles to broadcasters before shooting anything making the funding model in those days relatively reliable. Soon, however, the industry grew with the affordability and accessibility of prosumer equipment and editing software. The term "broadcast quality" entered the industry's lexicon to describe equipment previously used exclusively for home movies but now was sophisticated enough to produce video that met industry standards for television broadcast. The amateur documentarian was now competing alongside the professional for those broadcast license fees; so the industry responded by making production funding available only after the film was made instead of before production. The reasoning behind this was that if the quality of the finished product of the professional and the amateur is equal, we should simply look at the finished product of each and decide which one to buy. Everything being equal, the film with the most compelling content won the prize.

What separated the home-made documentary from the studio-made documentary was a term known as "unique access." The filmmaker was not as important as to what or to whom they had access and *unique* access was the ultimate goal. If a reclusive celebrity, for example, allowed just one person to interview them on camera, that filmmaker would almost certainly earn a broadcast license for the ensuing film.

When Michael Moore came on the scene with *Roger & Me* in 1989, he attracted a following of those interested in championing the cause of the "little guy." His aggressive and persistent search for General Motors CEO Roger Smith was seen as nothing short of heroic as Moore crusaded for the nearly 30,000 left unemployed—many Moore's friends and family in Flint, Michigan—and devastated by the closure of the GM plants there in the 1980s. His film, and his on-camera everyman persona, inspired a new generation of independent documentary filmmaker. When his next film—*Fahrenheit 9/11*—broke all box office records for the theatrically released feature-length documentary film in 2004, it ushered in a significant content adjustment between broadcasters and their audiences. The relatively light-hearted fare of Top 10 lists, celebrity biographies, and archival footage compilations had now taken a back seat to the much more popular social issue documentary. Following in the footsteps of Moore, the new documentarian was front and centre to audiences as much as their films. The filmmaker-activist often campaigned for change during festival screenings and long after their films left the cinemas.

Morgan Spurlock emerged in 2004 as a disciple of Moore, appearing in his own film *Super Size Me*, a takedown of the fast food industry and their profit-driven agenda that ignored their responsibility to their customers' health. Six weeks after the film's premiere, McDonald's dropped the "super size" option, yet denies the film influenced this decision. A food-related sub-genre soon grew in popularity with audiences, broadcasters, and activist filmmakers: *Food Inc.* (2008), *Dive!* (2009), *Forks over Knives* (2011), *Vegucated* (2011), *The Ghosts in Our Machine* (2013), *Fed Up* (2014), *Cowspiracy: The Sustainability Secret* (2014), and *Food Choices* (2016), to name just a few.

Several social issues were now being addressed by the activist documentary filmmaker, and film festivals soon followed programming, and in some cases dedicating their line-up specifically to social issue and activist documentary subjects. In particular, the growing global concern over environmental issues spawned many eco-specific film festivals. Here is a partial list of some of the environmental documentary film festival currently operating:

- Planet in Focus International Environmental Film Festival (Toronto, Canada)
- London Eco Film Festival (London, UK)
- Patagonia Eco Film Fest (Puerto Madryn, Argentina)
- Ecologico International Film Festival (Lecce, Italy)
- Environmental Film Festival (Melbourne, Australia)
- G2 Green Earth Festival (Venice, California)
- American Conservation Film Festival (Shepherdstown, Virginia)
- Environmental Film Festival in the Nation's Capital (Washington, DC)
- CMS VATAVARAN: International Environment and Wildlife Film Festival (New Delhi, India)
- Innsbruck Nature Film Festival (Austria)
- San Francisco Green Film Festival (USA)
- tiNai Ecofilm Festival (Chennai, India)
- Green Film Fest (Buenos Aires, Argentina)
- Princeton Environmental Film Festival (Princeton, New Jersey)
- Wildlife Conservation Film Festival (New York City)
- CinemAmbiente—Environmental Film Festival (Turin, Italy)
- Barcelona International Environmental Film Festival (Barcelona, Spain)
- Seoul Eco Film Festival (Seoul, South Korea)
- Finger Lakes Environmental Film Festival (Ithaca, New York)
- Colorado Environmental Film Festival (Golden, Colorado)
- Wild & Scenic Film Festival (Nevada City, California)
- Big Sky Documentary Film Festival (Missoula, Montana)
- Arctic Film Festival (Grand Marais, Minnesota)
- NaturVision Film Festival (Ludwigsburg, Germany)
- ECOCUP Green Documentary Film Festival (Moscow, Russia)
- Barents Ecology Film Festival (Petrozavodsk, Russia; and Joensuu, Finland)
- Paris International Environmental Film Festival (Paris, France)
- Kuala Lumpur Eco Film Festival (Kuala Lumpur, Malaysia)

Clearly, the urgency of addressing change related to how we treat the planet found a global audience, and the activist filmmaker was quick to supply documentary evidence of environmental issues and promote action among their audiences. Many documentaries with ecological themes such as climate change have a well-established and proclaimed agenda, and I don't mean to suggest this is an unacceptable approach by any means. Usually, these films are trying to correct a global crisis and focus on presenting the problem that few people may know and offering solutions from experts or even themselves in this progressive form of editorializing.

What I have seen happen when one of these films is complete is an active campaign on the part of the filmmaker and those individuals and organizations profiled in the documentary to reach out and mobilize the audiences that attend the initial screenings, usually at film festivals. Impact campaigns traditionally follow using social media to extend the audience reach and persistently encouraging the audiences to act by way of reaching out to the social issue's respective hegemony and demanding change.

As this book details, many times these campaigns succeed in motivating audiences to act, but often fail in clearly identifying their direct responsibility for causing a specific change in policy or law that corrects or eliminates the previously profiled problem, as Brian Winston et al. frequently indicate. When it does succeed—and these times are notably rare—I have noticed a consistent approach in place that significantly contributes to yielding change: when the audience is the changemaker.

The term "changemaker" is used throughout this book as a collective term for the individual or group who is ultimately responsible for and empowered to effect social change. This can be a policymaker, lawmaker, corporate CEO or board of directors, religious leader—anyone empowered to improve conditions in their respective communities by changing the rules, practices, or laws.

This book examines how and when this works by way of specific examples. *The Newfoundland Project*, produced by the National Film Board of Canada in 1967 (more commonly referred to as *The Fogo Island Film Project*); the films of the economically impoverished aboriginal Keewaytinook-Okimakanak communities in Northwestern Ontario made by George Ferreira between 2003 and 2005; *Sin by Silence*, Olivia Klaus' 2009 look at women imprisoned for defending themselves against domestic abusers; and my own films on climate change research at the ends of the world, *The Antarctica Challenge: A Global Warning* and *The Polar Explorer*.

What all these films have in common is that they are all seen as being representative of the rare time a direct connection between a documentary film and progressive social change can be made. I applied the academic method of performing a close reading on these films to see what they had in common, if anything, that might reveal itself to be a key ingredient in the recipe of producing social issue documentary films that have the ability of effecting social change. What I found lied not in the content or subject matter, not in the style or mode of production, but in the audience. These films succeeded when they were presented directly to the changemaker. When the audience of the general public was bypassed and the film was presented directly to the changemaker first, the goal of change sought by the filmmaker was achieved. This is not intended to suggest that this happens every time. There are, no doubt, several examples of when this approach failed, but in the rare instances when a direct link between a documentary film and social change is achieved, this approach is present. This technique is therefore evidence to me that a restructuring of the traditional creative and technical production model should be explored to identify a framework that might include the changemaker as a production collaborator.

My research into what this might look like began with an analysis of my own film projects in this regard. In 2008 I led a crew to Antarctica to produce a documentary about the climate research being done by the world's scientific community during International Polar Year (2007–2008). When I returned in 2009 and was completing post-production, I received a call from the Office of Communications of the United Nations Framework Convention on Climate Change (UNFCCC) inquiring about my film. It seems there was significant interest in presenting visible evidence of the impacts of climate change on a remote part of the world to the delegates attending the highly anticipated COP15 climate summit in Copenhagen that year, a conference dubbed by the media as "Hopenhagen."

I sent them a screener and they called me forty-eight hours later inviting me to come to Copenhagen and screen the film during the two-week conference. They indicated that my journalistic approach showing both the positive and negative impacts of climate change convinced them that this was a responsible, and not an incendiary, report of scientific research. The United Nations is careful not to champion overly emotional, extreme, clearly activist documentary films, even if their messages are factually accurate and represent necessary data in the communications process between scientists and policymakers. As ecocinema theory and practice were emerging in the late 1990s, the United Nations Educational, Scientific and Cultural Organization—more commonly known as UNESCO—issued a manifesto of sorts as a guideline for those engaged in communications endeavours related to urgent global environmental issues. The document titled *Educating for a Sustainable Future: A Transdisciplinary Vision for Concerted Action* was released in November 1997. The report eschews all forms of "alarmism" in communications products such as documentary film citing that "(e)xtreme positions, while they may be useful in catching the public's attention and alerting it to pending dangers, make it difficult to move from declarations and debate into action" (Subsection 53, Section II). Instead, the reports recommend that "(b)oth the messages and the messengers have to appear credible and responsible. Nothing is to be gained by scaring people. Alarmist predictions that make it seem that the world is about to end are evidently not conducive to the long-term planning and action that sustainable development requires. On the contrary, it is far more effective to present problems as manageable through responsible conduct and, wherever possible, put forward a realistic solution and a means to take preventive action" (Subsection 54, Section II). Since *The Antarctica Challenge: A Global Warning* met this criteria and satisfied the report's mandate for responsible communication, the United Nations Framework Convention on Climate Change made the bold proposal of including it as part of the conference programme, a "data delivery system" they called it that would be accessed by DVD, online links, and on-site screenings to delegates, negotiators, and world leaders throughout the two-week conference.

The UN's commitment to disseminating the film was impressive and swift. Thousands of DVD discs were made available to conference delegates in addition to daily inter-conference memos sent to all delegates sharing the film's secure link to screen the film online, and announcing the daily screenings and various venues throughout the Bella Center and other offsite locations in Copenhagen.

The screenings were numerous, twenty-five in total, with audiences of environment ministers, policymakers, world leaders, and NGOs. I was told the visual medium helped provide context to the written reports submitted by the global scientific community detailing their International Polar Year (IPY) research and findings.

The consistent journalistic approach to this documentary made it a data delivery system to the international environmental policymakers of the United Nations and made its filmmaker a witness, informant, and participant in the policy-creation process. Presenting the documentary directly to the changemaker sparked policy meetings and writing sessions as new data, now in a visible form rather than merely a written textual form, provided an additional context yielding a fuller understanding of the profiled issues.

Academic research supported this theory. In almost every case where there was a significant and measureable social change resulting from the impact of a documentary film, it was when that film had consulted with, or was screened specifically to, the changemaker. There are examples of documentaries that caused significant progressive change without involving the changemaker at any stage and relying solely on the public influence of general audiences who saw the film, but these examples are few. The one that most documentary film scholars point to is Gabriela Cowperthwaite's film *Blackfish* (2013) which examined SeaWorld's mistreatment of orca whales. Known as the "Blackfish Effect," the theme park nearly closed its doors permanently after public outcry of their handling of its sea creatures was exposed in the film.

Attendance numbers fell sharply in 2013, as did the park's market value, but as Aisha Hassan reports in a 2018 article for *Quartz Media*, the tide may have turned:

> The makings of a comeback were evident earlier in the year: Attendance in the quarter ending June 2018 was up 300,000, 4.8% more compared to the same period last year, and it reported an income of $22.7 million for the second quarter of 2018 compared to a net loss of $175.9 million in the second quarter of 2017. And despite no more breeding, the park will still have orcas in captivity for years, and SeaWorld is actively marketing them.[1]

The film certainly seems to have had an impact on SeaWorld's policy of animal treatment in captivity, but it seems to have survived the initial backlash and still keeps orcas in captivity as part of their revenue model. In this case, the film's audiences—many of them SeaWorld audiences as well—simply stopped going to SeaWorld and presumably influenced others to boycott the theme park as well. The point I wish to make is that a documentary film can, in fact, motivate an audience to take action, mobilizing the troops to storm the offices of the changemaker and demand change or

[1] Hassan, Aisha. "The 'Blackfish Effect' is over for SeaWorld—for now." *Quartz*, November 6, 2018. Accessed: July 16, 2019. https://qz.com/1451546/the-blackfish-effect-is-over-for-seaworld-for-now.

to boycott the offending corporation until change is made, but this surrogate groundswell does not seem to yield the same consistent results of measureable change as when the changemaker is consulted directly before, during, and after the filmmaking process.

I suspect no one likes to be told what to do, even when that which is being recommended is progressive and positive, but when the filmmaker treats the changemaker as a partner in the solution and not a target of the problem, the influence of a documentary is stronger and the impacts more immediate and measureable.

My first experience working with the changemaker as a production partner came in 2010 when I was asked by UNFCCC Director of Communications and Outreach Nick Nuttall to produce a film of Arctic climate research for the COP16 conference in Cancun, Mexico. Following a successful experiment in introducing the documentary film as a data delivery system in Copenhagen in 2009, a desire to repeat the process was suggested by Mr. Nuttall. He mentioned the importance of the journalistic approach and emphasized the value of footage related to glacier retreat study and evidence of new climate impacts in the Arctic. He also recommended an update of IPY research in Antarctica to be included in the new project.

As rising sea levels and their various causes were a focus of the Cancun conference, providing visible evidence of glacier collapse and in-field metrics of salinity decline and warming sea temperatures was going to be a valuable communications tool effectively bridging the gap between scientific evidence and policy creation.

The film made was called *The Polar Explorer* to reflect the research being done at both polar regions by today's explorer, the climate scientist. Our expedition crossed the Northwest Passage from west to east, conducting research at various stations along the way. As requested, we were able to provide the latest data on salinity levels, water temperature, new species of migrant marine life, the current location of the magnetic north pole, and, perhaps most importantly, footage of the Petermann Ice Island, a chunk of glacier ice that broke from the Petermann Glacier in Greenland just two months earlier. The film was also able to report data from Antarctica collected by a submarine that explored the underbelly of overhanging glacier tongues. I believe this marked the first time a film was produced and directed simultaneously in the Arctic and Antarctica, thanks to the digital affordances provided by online satellite telecommunications programmes like Skype.

The climate summit in Cancun took place from November 29 to December 10, 2010, leaving us very little time for post-production after we returned from our expedition in the Arctic in late October. We completed about 80 per cent of the film, including sound mix and scoring. As we had this unique premiere opportunity with the audience of changemakers that guided us in our content collection, we brought our cameras along to document the conference and the film's role in it. The footage we shot of screenings and meetings in Cancun made up the remaining 20 per cent of the film, resulting in a very meta depiction of a documentary film and its impact on change within the same film.

As this book details in Chap. 7, *The Polar Explorer* was directly responsible for the creation of a resolution that was included in the conference's Cancun Accord: *Enhanced Action on Adaptation: Section 2, Subsection 25.*[2] The process worked. The changemaker worked with the filmmaker before and during production, and the filmmaker worked with the changemaker after production at policy meetings providing supplementary information to the film and even contributing to the wording of the resolution.

This was all foreign ground to me and, I suspect, most documentary filmmakers. I served in this experimental capacity as a resource. The film provided the data and the filmmaker provided the context in which the data was collected. I was often asked to expand on such information as location of collected data, methodologies of data collection, and additional data available in reports written by the researchers profiled in the film.

It should be pointed out here that nowhere was the traditional audience of a documentary film seen. Between its production and its screenings in Mexico, there were no public exhibitions of the film anywhere. Having the general public act as agents of change, common to most social issue documentaries, was not a part of this production. The direct line connecting a documentary film with social change—so rarely identified—was clearly made in this collaborative process between changemaker and filmmaker. The general public was introduced to the film after the conference and after the scenes documenting the film's participation at the UN conference were included.

The Polar Explorer had its world premiere for the general public at the Canadian embassy in Washington, DC, in March 2011. Later that year,

[2] "Decision Adopted by the Conference of Parties," Cancun: United Nations Framework Convention on Climate Change, March 15, 2011. Web. Accessed: May 5, 2018. http://unfccc.int/resource/docs/2010/cop16/eng/07a01.pdf#page=4.

the film won twelve international film awards including the Audience Choice Award at the American Conservation Society's annual film festival and Best Experimental Film at the Alaska International Film Festival for combining a documentary film about climate change research with footage of that film's impact on international environmental policy.

Both the filmmaker and the changemaker were pleased with the experiment, and an investigation into how films like this can be made on a regular basis ensued. Since the UN does not finance external production, a different funding model needed to be created to accommodate the production of feature-length documentary films on climate change on an annual basis. The length of time to make a film combined with the sizeable budgets involved make it difficult under the best of conditions to produce an independent feature film once a year, so a different kind of film project designed to inform and influence power was conceived.

Mr. Nuttall expressed the UN's desire to involve youth more and provide them with a voice at the climate conferences. We developed an idea that would reach out to students worldwide to produce documentary shorts profiling climate research in their own countries. The five- to ten-minute films would be edited together anthology-style in a feature-length documentary film that would be screened to delegates attending the COP conferences each year. The same top-to-bottom approach of audience collaboration was used to inform the student filmmakers of the topics and kind of research that delegates needed to see for each conference.

The project was called the *Youth Climate Report*, and from 2011 to 2015, six such anthology documentary feature films were made and screened at COP17 to COP21, including one more for the Rio+20 Conference on Sustainability held in Rio de Janeiro from June 20–22, 2012. The films were presented at traditional screenings during the conferences, but since many delegates could not attend the screenings due to conflicts in their schedules during the two-week conferences, the films were made available online and links were circulated to all participating delegates. At some conferences, rows of wall monitors were set up to screen the film, with head-sets attached. Delegates could screen the film at their convenience in the halls of the conference centre.

While these efforts to accommodate screening schedules for the delegates were lauded, the same request kept coming back to us: to provide more content. The *Youth Climate Report* films showcased five to eight documentary shorts produced by students. This achieved the goal of involving youth in the UN climate summits, but the project's research reach needed to be extended.

As part of my course work in the PhD programme at York University's Department of Humanities, I took a course taught by new media scholar T.V. Reed, called *Digital Humanities*. It was in this course that I learned about the digital affordances of Geographic Information Systems (GIS) technology. GIS seemed uniquely structured to accommodate projects global in nature, and some of the existing software at the time could accommodate multimedia such as film. I was also inspired by work of Adrian Miles, a new media scholar from Australia who specialized in multilinear filmmaking. I soon envisioned a creative structure that united GIS software with multilinear storylines in a remediation of the documentary film that I call the Geo-Doc.

I worked with an open-source beta program Google was developing, called Fusion Tables. It was simple, but practical, for my purposes. I constructed a framework that would accommodate an infinite amount of student documentary films on climate research worldwide geo-locating each film to the coordinates of where the film was made. I would add to this metadata related to the year the film was made, its country, its latitude and longitude coordinates, links to written reports, climate researcher profiles, and institution websites related to that which is profiled in each film.

The UNFCCC Communications Office saw the potential to provide many more documentary film reports and assisted in a recruitment campaign issuing a call for entries to the media worldwide, encouraging youth participation. I used my own social media channels created for the project as well, and in our first year, we collected fifty films. The prototype—now called *The Youth Climate Report GIS Project*—was presented at the COP21 climate summit in Paris together with the last of our anthology documentary feature films: *Youth Climate Report 6*.

The project now collaborates with the Global Youth Video Competition, a programme designed to recruit videos from young people between the ages of eighteen and thirty. The filmmakers are given two to three themes per year to provide conference delegates with the latest information on those particular climate topics. The competition is a programme of Television for the Environment (TVE) and awards winners in each theme category a trip to the COP conference each year in whatever city is hosting. The top sixty films that make the short list in the competition are uploaded to *The Youth Climate Report GIS Project*'s database.

As the project grows in content (currently standing at 400 film units), so too does its ability to provide implicit narratives of time and space to the multilinear documentary film format. Each film unit in a geo-doc tells its own story, but a multilinear documentary film project becomes nothing more than a film library unless the film units relate to each other. Chapter 6 explains the importance of giving the user the choice to view the film fragments in the order they prefer, ostensibly making the viewer the editor; however, if the fragments are scenes or segments within a greater narrative, viewing the fragments out of sequence may result in story confusion for the user. Instead, if each fragment is a stand-alone mini-doc, the relation between each of them works when there is a common theme. In the case of the *Youth Climate Report GIS Project*, the common theme is climate research. In the case of another geo-doc project created for this book and examined in Chap. 6, *The Documentary Film World*, the common theme is social issue documentary feature films. The metadata contained within each pin provides not only support material to the corresponding film unit, but also data that has the potential to reveal implicit narratives when the pins are compared to each other. For example, film units in a specific area showcasing glacier retreat and separated by years could reveal a rate of decline to the user and thereby be used to project future rates of decline. In a spatial analysis, film units addressing the same glacier retreats in the Arctic compared to those in Antarctica might reveal climate impacts like air and water temperatures as they differ between the polar regions. The same temporal and spatial analyses are possible for any global social issue the geo-doc addresses, implicit narratives not always available in other multilinear documentary film projects.

It is important to note here that the strategy of collaboration between the filmmaker and changemaker throughout the production process does not yield a production exclusively crafted for policymakers. For any global social issue (poverty, violence, crime, women's rights, and especially all the related environmental issues: climate change, droughts, flooding, extinction, deforestation, pollution, the ozone hole, etc.), it is crucial that the geo-doc appeals to the general public as well as the changemaker. For this reason, it is recommended—but not essential—to use open-source GIS software in the creation of the production. The more contributions to the project that can be made by others will amplify the message and extend the educational outreach of the problems and proposed solutions to a global audience. Engagement with all audiences as contributors as well as spectators democratizes the process of the geo-doc production and endeavours to be as inclusive as possible in terms of voice, scientific data, and proposed solutions to provide as thoroughly representative presentation of the issue as possible.

We are moving in the right direction with respect to the current global climate crisis. In academia, the emergence and growth of studies in the environmental humanities, energy humanities, ecocriticism, and ecocinema are opening doors to new knowledge that aids in the informing of power and the discovery of potential solutions. But in all cases, information, proposed solutions, scientific data, and effective policy are not valuable without successful communication. Documentary film activism cannot achieve its goals while journeying down a one-way street. The information highway needs to accommodate two lanes of traffic for both the filmmaker and the changemaker.

Film projects like the geo-doc endeavour to do just that.

Toronto, ON, Canada — Mark Terry

Abstract

The documentary film has had a long history as an influential communications tool with the ability to effect social change. Its inherent claims to representing the truth provide a foundation of credibility that the filmmaker uses to inform and persuade their audience with a goal of causing them to take action that ideally leads to social change.

This goal has been seen to be achieved when the documentary film employs certain methods and technologies. My research questions are as follows: What methods and technologies are most effective in bolstering the documentary film's ability to effect social change and what new and emerging methods and technologies extend that ability? How can the documentary film be remediated to incorporate these attributes, and would this new project experience some measure of success in effecting social change when tested in the field?

These questions are answered through an investigation of various disciplines of study. The history of the documentary film as an instrument of social change is examined from its origins to the present day. This examination also identifies those methods and technologies that have advanced the documentary's ability to serve as a successful communications tool between filmmaker and changemaker. Focused investigations into the theory and practice of the documentary film yield specific approaches and techniques that prove to be most successful, such as the participatory mode, ecocinema and semiotic storytelling, the multilinear and the database documentary, and the distinct digital affordances provided by geomedia, all of which will be examined in this book.

Once identified and explained, the most effective theories and practices are combined in an altogether new and remediated documentary form: the Geo-Doc. This is a term I have applied to a structure of the documentary film that is a multilinear, interactive, database documentary film project presented on a platform of a Geographic Information System map. The project was made specifically for an audience of changemakers with the general public in mind as a secondary audience. In collaboration with the changemaker, content and interface suggestions are made to the filmmaker to augment the project's effectiveness as a communications tool.

A parallel investigation into the documentary's audience examines the effectiveness of collaborating with the audience before and during production as a means of enhancing a project's activist intentions. Instead of treating the audience as an end-user, the audience becomes a production partner. The content is driven, in part, by those who are empowered to effect social change, providing them with the specific content they identify as being required data for policy change. This concept reverses the gaze from the singular filmmaker to their audience of the general public, telling them, "take what I show you and demand change from those in power," to a collaboration between changemaker and filmmaker to produce a film that its audience of power says, "make a film with content I need to see and (social) change will occur."

To illustrate the effectiveness of this new documentary form and the proposed approach to changemaker collaboration, a case study using the global issue of climate change will be examined. The project, the *Youth Climate Report*, will demonstrate the geo-doc structure and its use in the field with the environmental policymakers of the United Nations.

Acknowledgements

Any book written is a collaborative effort. The author recognizes this contribution in the Acknowledgments section to thank publicly those who were instrumental in assisting in the research, support, and professional and artistic guidance from which every book benefits. Those listed below represent a team of contributors from a variety of fields that was essential in the creation of this book:

My Family: My patient, understanding, supportive, encouraging, and loving children: Herb and Mary Anne and their respective partners, Melissa and Ricardo; my wise and compassionate brothers Hebé (Herb) and Chip (Bob) and their respective wives, Miriam and Sharon, as well as all their children and grandchildren, too numerous to name here.

My Academic Colleagues: My supervisor, Gail Vanstone; my PhD committee members, Markus Reisenleitner and Seth Feldman; Sarwar Abdullah, Martin Bunch, Ravi de Costa, Sarah Flicker, Caitlin Fisher, Timothy Hudson, Helen Hughes, Susan Ingram, Adrian Ivakhiv, Brenda Longfellow, Janine Marchessault, Marcel Martel, Adrian Miles, Joan Steigerwald, Katherine Anderson, T.V. Reed, Chris Satoor, Erik Tate, Graham Wakefield, William Wicken, Brian Winston, James Orbinski, and the Dahdaleh Institute for Global Health Research. Special thanks to Lucy Batrouney, Bryony Burns, Mala Sanghera-Warren, and G. Nirmal Kumar of Palgrave Macmillan for their patience, assistance, and guidance through my inaugural book publishing process.

My Science Colleagues: Athena Dinar, Linda Capper, and the scientists of the British Antarctic Survey; Dr Martin Fortier and the scientists of

ArcticNet; Dr Yeugeny Karyagin and the scientists of the Ukrainian Akademik Antarctica; Dr Sylvia Earle and Dr David Ainley.

My Professional Colleagues: Bruce Cowley and the staff of the Canadian Broadcasting Corporation's digital channel, *documentary*; Nikolas Huelbusch and the staff at ZDF Enterprises; Amy R. Letourneau, David S. Matthews, and the staff of PBS Distribution; Anastasia Laukkanen, Bella Shakhmirza, Olga Dmitrienko, and the staff of the EcoCup Film Festival; Helga Stephenson and the staff of the Academy of Canadian Cinema and Television; Mark Madison, Chuck Dunkerly, Marianne Skye Elizabeth Tomasic, Karen Stoyko Fuegi, and the staff of the American Conservation Film Festival; Natasha Despotovic, Miryam López, and the staff of the Dominican Republic Environmental Film Festival; Jordana Aarons and the staff of the Planet in Focus Film Festival; Suzan Ayzcough, Damir Chytil, Daniel D'or, Alan Gibb, Philip Jackson, Carolyn Kelly, John Kelly, Michael Khashmanian, Ray Kocur, Holly Lee Rispin, Jaro Malanowski, Melanie Martyn, Janice Mennie, Tony Morrone, Jackie Pearce, Britta Schewe, Dianne Schwalm, Ken Simpson, and Stavros Stavrides.

My Explorer Colleagues: John Geiger and Joseph Frey of the Royal Canadian Geographical Society; George Kourounis, Maeva Gauthier, and Ginny Michaux of The Explorers Club; Lorie Karnath of The Explorers Museum; Tim Lavery of the World Explorers Bureau; the Russian Geographic Society, the Exploratorium, and the anonymous Member of Parliament from British Columbia who nominated me for the Queen's Diamond Jubilee Medal.

My United Nations Colleagues: Nick Nuttall, Catherine Beltrandi, Elisabeth Guilbaud-Cox, Fanina Kodre, Achim Steiner, and all the staff at the United Nations Environment Programme and the United Nations Climate Change secretariat; Tina Cobb, Laura Coates, Melita Kolundzic Stabile, Adriana Valenzuela Jimenez of the Education and Youth Office of the United Nations Climate Change secretariat; Cassie Flynn of the United Nations Development Programme; Charles Hopkins and Katrin Kohl of UNESCO; Minister Catherine McKenna, and all the members of the Canadian delegation to the United Nations Framework Convention on Climate Change; Nick Turner of Television for the Environment.

Contents

List of Figures

CHAPTER 1

Introduction

Act now! A familiar and effective plea in advertising, and as the name of a programme of the Crested Butte Film Festival, it is also the succinct message and goal many people associate with the documentary film. The tag line in their promotional poster[1] extends the description of this goal: "promoting social change through film"—another preconception many people share about the documentary film. But not all documentary films have activist intentions—biographies, nature films, travelogues, for example—yet the documentary film is often viewed as a genre of film privileged with the innate ability to serve as a powerful and influential communications tool, an instrument of social change. The ActNow programme description can easily substitute for an introductory definition of the social issue documentary and its activist intentions: "(the program/the documentary) inspires film-goers to become educated and respond immediately to certain films and social issues in a positive and proactive way, inviting social and environmental change."[2]

While few would argue this statement, it is important to remember that this is not the sole purpose of a documentary. A documentary with a social issue theme can also entertain (a mockumentary), deceive (propaganda), and train (educational films). The investigation of this book, however, will

[1] Poster: http://cbfilmfest.org/act-now-films.

[2] "ActNow," Crested Butte Film Festival, 2018. Link: http://cbfilmfest.org/act-now-films/.

M. Terry, *The Geo-Doc*, Palgrave Studies in Media and Environmental Communication,
https://doi.org/10.1007/978-3-030-32508-4_1

focus on the documentary's ability to communicate and influence for the purpose of achieving positive social change in a new way.

A fair question might be why attempt to change the documentary form in the first place. Does it require repair, or is it insufficient in its current state as an instrument of social change and is in need of an overhaul to better reach and influence those charged with creating policy that yields progressive social change? The answer is no, not entirely. The documentary film is enjoying a renaissance of sorts as many more people are using this communications device to speak truth to power. The documentary is not just for professional filmmakers any more. Amateur filmmakers, communications professionals, and even those in power themselves are taking advantage of the affordances and accessibility of digital technology to make more compelling arguments for their cause with the assistance of the visible evidence the medium provides. But the primary reason for this investigation and proposed reformation of the documentary is to explore news ways of advancing resolutions to our global battle against climate change.

Having served on the front lines of the Arctic and the Antarctica with the soldiers of science, I was able to see first-hand the devastating effects of climate change on the fragile polar regions. Participating in and documenting their research, I bore witness to new data essential in the surveillance of the enemy and in communication of this intel to headquarters—the international environmental policymakers of the United Nations. In this service, ecocinema is now playing an ever-increasing role, not only informing power, but all the other potential victims of climate change, the people of the world. Documentary activism is a powerful weapon, perhaps more now than ever before, but if there exists an opportunity to build a better weapon, it should be researched, tested in the field, and proposed for use. This is the goal of this book and its proposed new documentary form: the Geo-Doc.

* * *

The documentary film has long been associated with social change, and it has been used as an effective and influential tool by filmmakers, activists, and governments alike. But how does it do this? What techniques and technologies are most effective in bolstering the documentary and its claims to truth and effecting change? And once identified, do these specific approaches yield measurable impact, the goal of social issue documentaries?

Specifically, this book addresses the following questions. What methods and technologies are most effective in augmenting the documentary film's ability to effect social change and what new and emerging methods and technologies extend that ability? How can the documentary film be remediated[3] to incorporate these attributes, and would this new project experience some measure of success in effecting social change when tested in the field?

This book explores the various methods, approaches, and technologies that helped the documentary film develop into an ever-increasingly effective communications tool for social change agents and offers a new remediation of the genre that incorporates many of these proven techniques in a form that satisfies the aims and goals of both filmmaker and changemaker[4] in a new way. By incorporating the most evidently successful methods, styles, and digital affordances of the Internet as well as Geographic Information System (GIS) technologies, this book will introduce the GeoDoc, an altogether new form of the documentary film that hybridizes the proven successful styles and modes of activist filmmaking and the database format afforded by GIS technology to advance the film genre's contribution to effecting positive social change on a global scale. As well, I will investigate the relationship between the filmmaker and the changemaker as a strategy for maximizing the activist intentions of the documentary film. The binary of this relationship reveals an expedited path to social change when the changemaker-as-audience is engaged directly and when the changemaker collaborates directly with the filmmaker before and during production.

[3] "Remediate": The common definition is to remedy a situation, and while many will argue that the documentary is not in a state of disrepair requiring remedy, it will always benefit from enhancement; my use of the word "remediate" (and its various versions "remediating," "remediated," and "remediation") throughout this book focuses more on the part of the word "media." I propose a change to the traditional media of the documentary from a film viewed in a cinema or on television to a digital version as a multilinear project using Geographic Information Systems as a platform of exhibition and engagement, hence *reMEDIAting* the documentary film for the purpose of advancing its ability to inform and influence social change.

[4] "Changemaker": The individual or group ultimately responsible for making a change that benefits the society they represent, often a policymaker or lawmaker, but also a decision-maker, corporate head, board of directors, religious leader, or funding agency.

The first chapter examines the history of the non-fiction film and its emergent form as a documentary to provide a foundation of study with respect to how the documentary evolved over its years of use as an instrument of social change. Long believed to be a term coined by documentary pioneer John Grierson in 1926 (and cited as such by several documentary film scholars), the term, in English, was first used in 1907 by Charles Urban in a book he published to promote his educational films, called *The Cinematograph in Science, Education, and Matters of State*. The word, used as an adjective before it became commonly used as a noun, was first used in French in 1898 by Polish filmmaker and writer Bolesław Matuszewski. A review of these early days reveals experimental uses of non-fiction film for purposes of social improvement rather than mere entertainment.

The chapter chronicles the documentary film's early use to bring social reform in Canada, dating back to 1897. In particular, an investigation into this period reveals how the early film work of James S. Freer and Richard A. Hardie was instrumental in the government of Canada's promotional campaign to stimulate immigration and settlement in a new country. This sets the stage for the direct use by governments worldwide to exercise some measure of control over its citizens. Canada took the lead in this enterprise, establishing state-run film production studios regionally as early as 1914 in Ontario, in 1919 in British Columbia, and in 1920 in Quebec. These early experiments in government-produced documentary production led to national counterparts in the Canadian Government Motion Picture Bureau in 1923, and eventually the National Film Board of Canada in 1939.

Using the emerging documentary filmmaker as a tool of the state, the film genre and its claims to representing the truth are investigated to determine this period of discovery and how documentary pioneers began carving a new path for film, developing new modes of production, new goals for its filmmakers, and new ways of engaging its audiences. What methods worked and why, and on the flip side of that inquiry, what methods failed and why are questions addressed in this first chapter.

Chapter 2 explores these new advances by examining the numerous theories behind their creation and practice. One of the earliest practitioners of the community-based approach to documentary filmmaking—one in which the profiled community participates in the storytelling process—

was Joris Ivens, as far back as 1932. In his seminal canon of the life and work of this Dutch-born filmmaker, *The Conscience of Cinema: The Works of Joris Ivens 1912–1989*, Thomas Waugh details Ivens' achievements with this approach with such films as *New Earth* (1932), *Misère au Borinage* (1934), *The Spanish Earth* (1937), and *The 400 Million* (1939).

A key issue debated by documentary theorists is the film genre's claim to representing the truth. John Grierson's famous definition of the documentary as a "creative treatment of actuality" is contested by many scholars. They point to the word "creative" as giving licence to the filmmaker to distort or misrepresent the truth in the name of art. This theoretical investigation targets such classic documentary films as *Nanook of the North*, revealing scenes that were staged by its filmmaker Robert Flaherty who once famously defended his actions by stating, "Sometimes you have to lie in order to tell the truth."

Cognizant of this contentious issue, Russian filmmaker Dziga Vertov adopts an absolutist approach to representing the truth in his documentaries by describing them as *kinopravda* (film truth). According to documentary film scholar Bill Nichols, Vertov "eschewed all forms of scripting, staging, acting, or re-enacting," choosing instead to represent reality as the camera captures it without any literary or theatrical structure. Perspectives on this aspect of Vertov's work are provided to investigate the contribution Vertov made to non-fiction film and his purist approach to representing the truth in documentary projects as an essential strategy to effect social change.

The audience is another focus of this chapter on documentary theory. The key target of audience engagement for documentary filmmakers is often identified as emotion. It is argued that if the film and its messages can reach an audience emotionally, it is likely they can then be moved to take the action the filmmaker wants them to take. In providing context to this aim of the documentary, Michael Renov provides four functions of the documentary film: "to record, reveal, or preserve; to persuade or promote; to analyze or interrogate; to express." Through an examination of these theories, the documentary's other roles as a communications device are identified and explored.

Related to this is Bill Nichols' theory on the seven modes of documentary filmmaking: participatory, expository, poetic, observational, reflexive, performative, and interactive. These approaches, Nichols argues, yield different results in their audience engagement and often have goals different from effecting social change, but the mode that seems to be used the most for this specific goal is the participatory mode. Examples of the effective use of this mode are examined through the National Film Board of Canada's *Newfoundland Project* (more commonly known as the *Fogo Films*), the films of Father Albert Tessier, and the films of George Ferreira, all documentary projects that not only involved their interview subjects in the filmmaking process, but also engaged its audience of changemakers directly.

Is the participatory mode essential for the documentary film to achieve its activist goals of positive social change? Is involving the interview subjects enough, or should participation extend to the changemaker before, during, and after production? I posit that a more comprehensive and inclusive participation take place with all stakeholders—including the audience of changemakers—at all stages of production to expand the documentary's ability to contribute to social change and to increase its rate of success in doing so.

A relatively new area of documentary theory is explored in Chap. 3: ecocinema. A term first coined by film critic Scott MacDonald in his 2004 essay "Toward an Eco-Cinema" [sic],[5] ecocinema, as a sub-genre of documentary, employs semiotic storytelling techniques and implicit narratives to enhance its ability to influence audiences on environmental issues. MacDonald specifically addresses the use of these techniques in ecocinema to "retrain our perception," a sentiment first expressed by eco-philosophical theorist Felix Guattari almost ten years earlier. In his essay "Chaosmosis: An Ethico-aesthetic Paradigm," Guattari calls for a "mutation of mentality" to understand and find solutions to environmental problems.

This cinematic ecosophy declares this approach is a more effective way of moving audiences to action than simply appealing to them emotionally. As examples of this technique, this chapter explores the work of DJ Spooky (*Sinfonia Antarctica*), my own 2009 eco-doc (*The Antarctica Challenge: A Global Warning*) and the History Channel's eco-cinematic series *Life After People*. In all three examples, the theory of ecocinema is employed. Specifically, semiotic storytelling techniques underscore explicit narratives to create a process-relational model. This model combines the perceived world on the screen with the actual world off screen, providing the audience with a more profound and relatable understanding of the environmental issue depicted in the film and amplifying its call to action.

[5] After this first paper, MacDonald changed the spelling of the term he coined from "eco-cinema" to "ecocinema."

It is no coincidence that the dawn of ecocinema coincides with the advent of digital documentary storytelling. The unique affordances provided within the digital domain give application to theories like those employed in ecocinema. In Chap. 4, the revolution of the digital documentary is explored to identify the areas in which digital technology has replaced traditional documentary techniques, media, and approaches and practices, including audience engagement. Michael C. Nisbet and Patricia Aufderheide identify the documentary's newly evolved purpose of social activism in the digital domain:

> Documentaries are no longer conventionally perceived as a passive experience intended solely for informal learning or entertainment. Instead, with ever increasing frequency, these films are considered part of a larger effort to spark debate, mould public opinion, shape policy, and build activist networks. (Nisbet, Aufderheide, 450)

This is achieved through a variety of digital affordances that have refined the documentary with respect to filming, editing, distributing, exhibiting, and audience engagement. In each case, there exists some measure of enhancement to the documentary's goal of influencing audiences. However, there are dissenting views that insist the digital world is a place that can overcomplicate the documentary process, providing too many options and processes which ultimately result in audience disengagement. This book contends that this objection can be recognized and controlled in a manner that does not confuse the audience, but rather supports its desire for a fuller understanding of the profiled issue.

In particular, this chapter examines how the digital version of the participatory mode, inclusive collaboration, extends its reach to audiences of the general public and those charged with creating social change. In one example—the documentary film *The Price We Pay*—the policymaker herself sought out the professional documentary filmmaker to create a film that told the story of off-shore tax evasion so that she could use the film as a communications tool to introduce new legislation to stop the practice. In another example, a comparison is made between Colin Low's successful attempt at community filmmaking, commonly known as *The Fogo Process.* Its digital counterpart, the films of George Ferreira, is also examined. Ferreira's documentaries were designed to provide visible evidence to the federal government of Canada of the economic plight of the aboriginal Keewaytinook-Okimakanak communities in Northwestern Ontario. These films also reveal the extended positive results yielded by the digital affordances—and cost-saving measures—of live streaming.

This leads to a discussion of the collaborative documentary. Many of the same techniques and approaches are employed in the creation of these films, but the collaborative opportunities afforded by the digital domain make this process more democratic, often involving filmmaker, interview subject, changemaker, and public audiences simultaneously. While Kate Nash believes there is no concrete evidence "between interactivity and audience empowerment," Jon Dovey argues that the two work together as essential elements in a dynamic documentary ecology. He illustrates this relationship by categorizing key components in the digital documentary: affiliation, expression, collaborative problem-solving, and circulations. This creative ecosystem may be attractive and even effective to social change filmmakers, but in its infancy, some problems exist, as film theorist Sandra Gaudenzi (2017) suggests in her essay "User Experience Versus Author Experience":

> One could make a case that it is just a question of time, and that changing workflows in teams and institutions is a slow process. One could also point out that interviewing users and testing prototypes does involve a cost, and that small productions cannot afford the process. To these two valid explanations I would like to add a third one: that for many documentary makers there is resistance to the idea of testing their work. (Gaudenzi, "User Experience," 124)

One of the examples of a successful digital documentary project is the 2009 film *Sin by Silence*, made by Olivia Klaus. Collaborating directly with audiences of the state as well as prisoners convicted of murdering their domestic abusers, Klaus was successful in reopening investigations and, in one case, having the sentence over-turned and the prisoner released. The connection to the film as the instrument of this change was attributed directly by Governor Jerry Brown of California. Even the bills that were passed into law bore the name of the film: *The Sin by Silence Bills.*

This is a clear example of a direct connection to a documentary film and its ability to influence the changemaker, but these cases are rare. To steer the documentary film in the direction of creating social change and to provide a measurement of success when it does occur, the digital domain has given rise to several approaches, programmes, and technologies designed to achieve social change and identify when it does. Organizations such as BRITDOC, whose motto is "Dedicated to the Impact of Art and the Art of Impact," provide filmmakers with a set of digital tools to assist in their social justice filmmaking. This section of Chap. 4 focuses on such tools as *ConText*, *Harvis*, *Media Cloud*, and *Ovee*, all designed to engage and measure audience action and reaction.

Conceptually, one of the biggest changes in the digital revolution of the documentary film is the way it can now be made and presented. Entirely new methods of production, dissemination, and exhibition have changed the essence of the documentary, and Chap. 5 focuses on this. Extending the theory of the participatory mode and the collaborative, community-based filmmaking process, the multilinear documentary offers a single platform expressing the viewpoints of many voices on a single issue. Theoretically, it is a democratized approach to the social issue documentary, and while some flaws exist, I argue that the overall contribution to influencing the changemaker is heightened by providing multiple voices, perspectives, and narratives which collectively yield a fuller understanding of the social issue to the changemaker.

Just as ecocinema places emphasis on the relation between the world depicted on the screen and the actual world off the screen, multilinear theory stresses the importance of inter-relationality between its film fragments. It is important to note that the multilinear documentary works only when the individual film components *relate* to each other. The Korsakow System (a software enabling the filmmaker to create projects with multiple film fragments presented simultaneously), one of the first multilinear experiments, has been criticized for failing to do this, making the overall viewing experience *flat* or *static*. The absence of a fixed editorial structure results in an audience with no emotional engagement.

This is valid criticism, but one that can be avoided by employing more of the affordances of the web and by ensuring the film fragments contained within the project maintain relationality to each other. In addition to content breeding context, additional information, or metadata, can be provided to assist in making the film units relate to each other. One of the platforms that can provide this structure is the Geographic Information System (GIS) map. In this chapter, I argue that additional perspectives such as time and space are afforded by GIS platforms, thereby providing the relationality that the film components of a multilinear documentary project require: "Within a GIS users can discover relationships that make a complex world more immediately understandable by visually detecting spatial (and temporal) patterns that remain hidden in texts" (Bodenhamer et al. 2010, vii). From what we learn in Chap. 4 about ecocinema, this semiotic affordance of GIS provides a complementary construct that lends itself well to an environmentally themed documentary project on a global scale.

To this end, I have created a geomedia project incorporating the proven techniques, approaches, theories, structures, and concepts of the documentary film, examined in the previous chapters, and incorporated them in a GIS platform on the global environmental issue of climate change. The project is called the *Youth Climate Report GIS Project* (YCR), and it was originally designed using a beta GIS technology created by Google, called Fusion Tables.[6] The project is a multilinear, interactive, database documentary film project on international climate change research created collaboratively with the global communities of youth, science, and the environmental policymakers of the United Nations. In addition to more than 400 documentary short films, the project incorporates the dimensions of both time and space as well as metadata such as written texts, photography, and web-links to provide the relational context required by multilinear projects and to engage audiences on a variety of levels of understanding.

The project is detailed and examined critically in the final chapter of this book. As a case study, it builds on those aspects of the social issue documentary that have proven to yield successful change in the preceding chapters. To test the success of this approach, the YCR project was introduced to the United Nations as a new data delivery system for delegates attending COP21, the framework convention on climate change held in Paris in 2015. Engagement with the prototype was encouraging, and following the addition of more films, as requested by the UN delegates, the project was officially adopted by the UN under its Article 6's mandate for education and outreach at the 2016 climate summit in Marrakech, Morocco.

I call this remediated form of documentary the Geo-Doc and, as such, can be used to present any social issue, not just environmental ones, in collaboration with and directly to the changemaker. The geo-doc has the capacity to expedite the process of influencing social change by providing a fuller understanding of the issue to the changemaker with metadata and

[6] As Google has decided to discontinue *Fusion Tables*, the *Youth Climate Report GIS Project* has migrated to a new open-source platform offered by Google, known as *My Maps*. The new technology offers improvements to Fusion Tables such as extended affordances that enhance the project's interface and presentation of information contained within the film units.

by providing a voice of those impacted by the issue, so they may speak directly to power. It can also be directly and immediately sent to the changemakers by way of a link—no cinema, television or video player is required, just a computer or even a simple mobile device.

Naturally, there is still room for improvement. The geo-doc could also benefit from other digital affordances such as a live-cam component, so that audiences can see the area or engage with the people profiled in the documentary project in real time. Filmmakers will also undergo a transformation of sorts as their role as auteur changes to more of a curator. In addition to collecting film units made by other filmmakers to populate a GIS map, the geo-doc filmmaker will also be collecting written reports, photographs, web-links, and other uploadable data to enhance each film unit, and in so doing, the overall project as well.

This new platform for the documentary may never be screened at the Crested Butte Film Festival, but it may very well augment the documentary's goal of influencing audiences to "Act Now."

Bibliography

Bodenhamer, David J., John Corrigan, and Trevor M. Harris, eds. *The Spatial Humanities: GIS and the Future of Humanities Scholarship*. Bloomington, IN: Indiana University Press, 2010. Print.

Gaudenzi, Sandra. "User Experience Versus Author Experience." In *i-docs: The Evolving Practices of Interactive Documentary*, ed. Judith Aston, Sandra Gaudenzi and Mandy Rose. New York: Columbia University Press, 2017. Print.

Nichols, Bill. *Introduction to Documentary*, 3rd ed. Bloomington, IN: University of Indiana Press, 2017. Print.

CHAPTER 2

Farming the Tools of Persuasion

2.1 Introduction

Confucius once famously said, "Study the past if you would define the future" (Zuang, Baker, 24). Following this advice, this chapter explores the history of film to trace the path that led to the genre of documentary from its raw origins as non-fiction. This important first step provides the foundation of understanding how the documentary film came to speak to power as an influential tool of social change, a foundation that filmmakers and related stakeholders continue to build on with new technologies, new modes of storytelling, and new platforms of exhibition. This also provides a foundation that new media theorists use to expand on to develop new approaches that subsequent chapters will examine. As the history of film and the documentary is extensive, this chapter will focus on only those years (the late nineteenth century to the mid-twentieth century) and events that were formative in terms of how non-fiction film developed in Canada as the social issue documentary and as an effective and influential communications tool.

As a form of entertainment, the moving picture was an enhancement of photography, a technique that attracted many to its unique ability to capture reality within a single frame of film. It can be argued that the photograph itself was an enhancement, as well, of the painting whose ambitions were "the expression of spiritual reality wherein the symbol transcended its model" and to "duplicate the world outside," according to Andre

M. Terry, *The Geo-Doc*, Palgrave Studies in Media and Environmental Communication,
https://doi.org/10.1007/978-3-030-32508-4_2

Bazin in his essay "The Ontology of the Photographic Image" (Bazin, 4–9). But neither the painting nor the photograph exhibited the power that film grew to yield in influencing those in positions of implementing social change.

The dimension of movement to capture moments of reality demonstrated a deeper engagement with viewers and one that provoked emotional responses among its viewers. The discovery of this ability was in large part unintentional, even accidental, but once recognized, control over the film content to control audience's beliefs and behaviour became a focus of governments, educators, social change activists, and profiteers alike.

2.2 The Early Years

In the nineteenth century, the art and science of photography opened a door to the emergence of many new experimental inventions to capture reality as permanent memory images that would eventually lead to the birth of the genre of film we now know as the documentary, but its aesthetic origins can be traced back even further.

The "Magic Lantern," developed in the 1650s, was a crude projector of images using a lens to enlarge pictures and simple candlelight to illuminate them. The candle was replaced by the Argand lamp in the 1790s, followed by the invention of the Drummond light—more commonly known as "limelight"—both enhancing the brightness of the projected images (Waddington, xiii–xv). This technology primarily served the purpose of entertainment during this 140-year period.

In 1823, French painter Louis Jacques Mandé Daguerre was responsible for two inventions that advanced the emergence of the motion picture: the diorama (an early cinema) and the self-named daguerreotype (a photographic process). By using a camera obscura, scenes viewed by the device would be captured on a silver-plated copper plate and projected in panoramic fashion against a wall. In 1839, this photographic process was perfected and introduced in Paris, heralding the birth of photography ("A Daguerreotype of Daguerre").

For the first time, images of "reality"—as opposed to artists' interpretations of reality—were being preserved and presented for mass consumption. Using these projected photographs, other inventors soon developed rudimentary moving pictures. Revolving drums, with a series of photographs taken in quick succession placed within them, would give the illusion of motion to the observer. These were developed by Simon von Stampfer (Stroboscope) in Austria, Joseph Plateau (Phenakistoscope) in Belgium, and William Horner (Zoetrope) in Britain ("History of Film Technology").

The first motion picture photographed in real time was created in the United States in 1878 by British photographer Eadweard Muybridge to determine whether a galloping horse named Sallie Gardner ever had all four feet off the ground at the same time. Several cameras were set up to capture this motion. Each camera shutter was triggered by a thread as the horse passed and each exposure was made in only one-thousandth of a second (Clegg, 129). The significance of this is that the goal of creating moving pictures was to demonstrate a reality invisible to the human eye, but visible to the mechanical eye. This was considered more trustworthy in representing the truth, an inherent ability that many have attributed to non-fiction film—a claim this book will examine more closely in subsequent chapters.

2.3 1895 to 1903

Other inventors refined this technique, but it was Auguste and Louis Lumière who introduced the world to their specific new technology of moving pictures at a public exhibition in Paris at the Salon Indien du Grand Café on December 28, 1895, showcasing ten short films of about fifty seconds each in duration (*Les Films Lumière*). These early films included *La Sortie des Usines Lumière à Lyon* (*Workers Leaving the Lumière Factory in Lyon*).

One of the first film reviews outside of France came from Russian writer Maxim Gorky in 1896. In reviewing the Lumière films, his critique focuses more on the style of the film, rather than on its content, describing the viewing experience as watching a pale imitation of real life:

> Their smiles are lifeless, even though their movements are full of living energy and are so swift as to be almost imperceptible. Their laughter is soundless although you see the muscles contracting in their grey faces. Before you a life is surging, a life deprived of words and shorn of the living spectrum of colours—the grey, the soundless, the bleak and dismal life. (Macdonald, Cousins, 13)

Other newspaper reports of the Lumière screenings of 1895 also addressed the technology more than the content of the films. The Parisian newspaper *La Radical* wrote:

> Already, words are collected and reproduced; now life is collected and reproduced. We can, for example, see our dear ones again long after they would be lost to us. (Matsuda, 172)

The two reviews are significant in that they represent both sides of the debate between whether or not documentary film depicts the truth. While Gorky admits to witnessing "life" in the people presented in the films, he describes that life as "bleak" and "dismal," lacking the components we take for granted in "real life": colour and sound. The result is a reality truer than a static photograph, but still less true than reality itself. The description in the report in *La Radical*, on the other hand, heralds the captured "life," extending it beyond the death of those depicted in the film, attributing a "truth" to the images that would not be questioned by even the "dear ones" witnessing these filmed lives.

Often considered the world's first documentaries in terms of representing reality instead of fictive stories, the Lumière brothers used the term "Actualités" as a descriptor in the printed catalogues of their films (*L'histoire Lumière*). These non-fiction films were not yet documentaries as we have come to know the genre; however, while early audiences were very likely convinced that the workers leaving the Lumière Factory were some of the finest dressed factory workers in France at the time, we now know this is not true from some of the other versions of this film. There are at least three versions—or takes—of this scene. The first one is the closest one to depicting the probable reality of the time, as the workers are seen to be wearing clothing more typical of those working in a factory. The other two versions show the workers dressed in fancy finery.[1]

Perhaps the first instance of mistrust of the non-fiction film, the questions about the authenticity of the content of documentaries arise now when we see certain aspects altered to create different meaning: were the well-dressed factory workers wealthy, happy, and not working in a messy environment as presented in one take, or were they the blue-collar labourers wearing work clothes as presented in another take? Each version tells a

[1] All three versions of *La Sortie des Usines Lumière à Lyon* can be found on YouTube: https://youtu.be/DEQeIRLxaM4.

different story about the working conditions at the factory and about the Lumière brothers as employers; which one represents the "truth"?

While shining a light on the employees of the Lumière factory, the working class was given a face to all of France and the world. This promising new technology began to demonstrate its utility as a democratizing device, engaging and illuminating the working class while simultaneously entertaining all classes. This early association with elevating the common man—or at least not excluding him—and making powerful new technology available to all, paved the way for film as a new medium to promote positive social change. This goal runs consistently throughout the documentary film's development despite shifting perspectives from its artists, commissioners, exhibitors, and producers.

This egalitarian shift in the art world of Western civilization at this time was not lost on the Lumière brothers when they chose to showcase their working-class employees in one of their first films. At the same time, they also chose to film another form of democratizing the classes, the train, a mode of transportation for all, in a film that demonstrated film's substantive power to move audiences emotionally: *L'Arrivée d'un train en gare de la Ciotat* (*The Arrival of a Train at the Ciotat Station*), first screened on January 26, 1896 (Loiperdinger and Elzer, 102).

When the Lumière brothers chose to film a train, they created an experience that moved the audience like never before. It has been reported that those watching this film for the first time grew increasingly agitated when they saw the cinematic train "coming at them" as it entered the Ciotat Station. The extreme reaction to this early film provides evidence of the power of film to move audiences emotionally, an attribute necessary to move viewers to action and cause social change in the years to come, which we will investigate more deeply later.

The Lumière films had a tremendous influence in developing an emerging documentary film industry. Moreover, their particular content seemed to inspire one Canadian documentary pioneer in particular and launch a journey that would lead to the world's first state-run film studio, a development that represents one of the first applications of the documentary film as an instrument of social change.

2.4 Canada, 1897

Two years after the historic Lumière screenings of 1895, a former British journalist born in Woodstock, Oxfordshire, on January 4, 1855, now a Canadian farmer living in Brandon Hills, Manitoba, James Simmons

Freer, bought himself one of Edison's film cameras and became "Canada's first filmmaker" (Morris et al. 2012). Unintentionally, Freer produced Canada's first documentary films when he filmed his farm and the local train passing by. One of his first films was titled *Arrival of CPR Express at Winnipeg* (1897). Not only was the title almost identical to the Lumière brothers' *Arrival of a Train at the Ciotat Station*, but so was the content, almost frame-by-frame, as the short film depicted a train in the distance approaching the camera and its filmed audience in spectacular cinematic fashion. It is not known how long this film was, but one Canadian film historian, Gene Walz, believes it to be "not longer than two minutes" (Walz, 3).

Around the same time, another individual on a tour across Canada was promoting locally produced films, thus casting doubt on Morris' claim that Freer was "Canada's first filmmaker." Richard A. Hardie, described as a "showman" by film historian Paul Moore, "embraced nationalism to produce Manitoba booster films, intended to promote immigration from the United Kingdom" (Moore, 74). Moore goes on to detail a Canadian tour of locally produced films managed by Hardie and competing showman William McCarthy in May 1897 (Moore, 84), six months before Freer's tour in the UK.

In a period representing about five years (1897 to 1902), Freer produced numerous non-fiction films showcasing Canada. The film titles included *Arrival of CPR Express at Winnipeg*, *Pacific and Atlantic Mail Trains*, *Harnessing the Virgin Prairie*, *Canadian Continental Jubilee*, *Premier Greenway Stooking Grain*, *Six Binders at Work in Hundred Acre Wheatfield*, *Typical Stooking Scene*, *Harvesting Scene with Trains Passing By*, *Cyclone Thresher at Work*, *Coming thro' the Rye* (a film depicting children playing in the hay), *Fire Boys on the Warpath*, *Canadian Militia Charging Fortified Wall* (Morris, 30). Unfortunately, none of Freer's films have survived (Frank, 26). Were these the same films with which Hardie and McCarthy toured Canada, and if they were, who made them, Hardie, McCarthy, or Freer? Titles of these pre-UK tour films are not available, but descriptions of them are: "the Winnipeg and Brandon fire brigades racing down city streets, sidewalk crowds, trains racing toward the camera, and plenty of wheat being harvested" (Moore, 85). In fact, Moore believes "these are the same films Freer brought to England" and surmises that "(a)lthough Freer has long been assumed the filmmaker, he did not travel abroad until December 1897, months after the films had already been exhibited across the Prairies" (Moore, 85).

As evident in some of the titles, Freer's work became more and more linked to government commissions rather than the arbitrary subjects of farm and family afforded to a truly independent artist. He was becoming a pawn of the state, and he would soon be used by the federal government for the purpose of promoting immigration overseas.

Collectively, these films were known as Freer's *Ten Years in Manitoba* film series (Morris, 30). His promotional expedition represented the first on-location shoot for a Canadian filmmaker as he made the films *Canadian Contingent at the Jubilee* and *Changing Guards at St. James Palace* while touring England (Morris, 30).

Taking his films throughout England with the financial support of one of the stars of his films, the Canadian Pacific Railway (CPR), significant interest in moving to Canada was produced because his films showed "the value of agricultural pursuits in Canada" (Walz, 3). That value was identified as rich, fertile soil and "large free grants of land" (Walz, 1 and 9) that the Canadian government was offering to immigrants at the turn of the century.

This tour represented the first partnership in a film project between filmmaker (Freer) and the corporate sector (CPR). Each participant had a goal to achieve with this tour: Freer would advance his skills as a journalist with this new technology of filmmaking, and return home to friends and family as a Canadian film pioneer; the CPR, a great believer in modern promotional methods at the time, would use the filmmaker and his films as promotional tools in its land settlement policies and to attract customers (Morris, 30). The anticipation of Freer's screenings in England was high. Printed flyers were posted throughout London, advertising the historic event in cinema history (see Fig. 2.1). The significance of this enthusiasm shows the credibility given to Freer and the excitement for moving pictures before his films were even shown.

Freer's UK screenings to friends and family and to the general public were a great success. Perhaps the first film review of a Canadian film came from the *London Daily Mail* reporting on Freer's screenings. The newspaper described Freer's films as a "capital series of cinematograph pictures," while the *Eastern Daily Press* described them as "reproducing in realistic manner the conditions of life in the Far West from the interior of a bachelor's shanty to Mr. Freer's pretty and attractive family residence with the family assembled outside" (Morris, 30–32). Perhaps the most telling of the media coverage his films received in England was how the motion pictures were described as an "emigration agent" (Morris, 30). This indicates

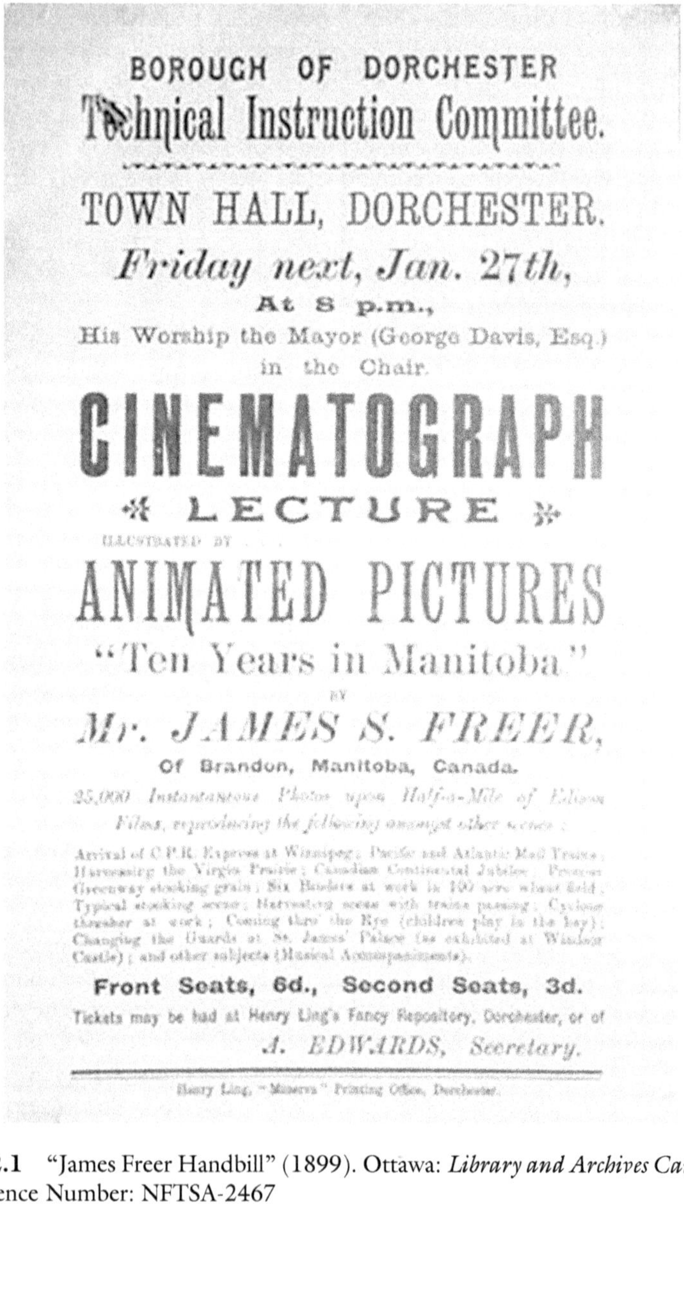

BOROUGH OF DORCHESTER

Technical Instruction Committee.

TOWN HALL, DORCHESTER.

Friday next, Jan. 27th,

At 8 p.m.,

His Worship the Mayor (George Davis, Esq.)
in the Chair.

CINEMATOGRAPH

LECTURE

ILLUSTRATED BY

ANIMATED PICTURES

"Ten Years in Manitoba"

BY

Mr. JAMES S. FREER,

Of Brandon, Manitoba, Canada.

25,000 Instantaneous Photos upon Half-a-Mile of Edison Films, reproducing the following amongst other scenes:

Arrival of C.P.R. Express at Winnipeg; Pacific and Atlantic Mail Trains; Harvesting the Virgin Prairie; Canadian Continental Jubilee; Premier Greenway stooking grain; Six Binders at work in 100 acre wheat field; Typical stooking scene; Harvesting scene with trains passing; Cyclone thresher at work; Coming thro' the Rye (children play in the hay); Changing the Guards at St. James' Palace (as exhibited at Windsor Castle); and other subjects (Musical Accompaniments).

Front Seats, 6d., Second Seats, 3d.

Tickets may be had at Henry Ling's Fancy Repository, Dorchester, or of

A. EDWARDS, Secretary.

Henry Ling, "Minerva" Printing Office, Dorchester.

Fig. 2.1 "James Freer Handbill" (1899). Ottawa: *Library and Archives Canada.* Reference Number: NFTSA-2467

the first time the filmmaker was acknowledged not as an artist, but as a public servant—or tool—of the state.

Since Freer was Canada's first filmmaker (as opposed to film showman), this also marked one of the early moments in the history of Canadian non-fiction films that were seen to serve a purpose other than entertainment. These London newspapers saw Freer's films as promotional pieces for moving to Canada. Although not originally intended as such, they did, in fact, serve that purpose in the end (Morris, 30). James Freer started out as an artist or, perhaps more accurately, as a journalist, documenting his observations with a motion picture camera instead of a pencil and paper. The Government of Canada was now using his filmed reports—and Freer himself—for the purposes of immigration propaganda. Perhaps believing they were pursuing the altruistic goal of serving the public good, the Canadian government was intent on attracting immigrants to fulfill its own need for population expansion to protect the Canadian border against American incursion.

The success of this tour did not go unnoticed by the Canadian government. Freer's neighbour and friend, Clifford Sifton, Federal Minister of the Interior at the time, enthusiastically threw the weight of the federal government behind Freer's productions and British tours for the purpose of stimulating immigration to Canada. Furthermore, Sifton also reached out to Ukrainians during this time, attracting 170,000 immigrants to the Prairie provinces (Gagnon 2015).

> Sifton had close ties to the Canadian business community, and he attempted to apply new techniques (i.e. film) of business promotion, management, and advertising to the task of attracting settlers ... the department enjoyed its greatest success in Continental Europe, where immigrants streamed from the farming regions of Poland, the Ukraine, Germany, and Austria-Hungary. ("*Immigrant Voice*")

It is not known if Sifton used Freer's films as promotional tools for the Ukrainians, but since he was using them for his British immigration campaign at the same time, it is highly likely that he used them to promote Ukrainian immigration as well between 1898 and 1905 (Gagnon 2015). The Toronto Ukrainian Genealogy Group offers this:

> Although eighty percent of the Ukraine's land was part of the Russian empire, most Ukrainians that landed on Canada's shores during this period (1897–1917) were from the western portion, then under Austrian rule. The

> first immigration consisted almost entirely of land-hungry peasants from the provinces of Galicia and Bukovina. ("*Ukrainians*")

Denied any opportunities to improve their lot in their homeland, these immigrants were attracted to Canada by its policy of granting virtually free lands or "homesteads" to settlers (*Ukrainians*). This was the same promise made by Freer to his UK audiences.

At this time, the population of Canada was five million. Between 1897 and 1917, through the combined efforts of Freer's films, the CPR and the federal government, three million immigrants came to Canada in this twenty-year period. It is believed by historian Peter Morris that Freer's films played a key role in this "unprecedented process of persuasion" (Morris, 33). George Melnyk echoes this in his analysis of the contribution made by Freer's films:

> Freer's films became part of the Canadian Pacific Railway's and the Canadian government's push to attract settlers to the West when Canada had only a population of five million. As a result of this intense drive for settlers, the population increased by another three million. (Melnyk, 15)

According to historian James H. Gray, the Canadian government severely underestimated its own promotional efforts which included film for the first time. The huge wave of immigration arrived at a land "where not a single constructive step had been taken by anybody to prepare for their arrival" (Gray, 15). The Ukrainian and British people who swarmed and settled in Western Canada provided significant evidence of film's ability to bring about social change. They also represented the first case of a country's government getting involved in the new art form of film, a partnership that would come to define Canadian cinema in the years to come.

As previously indicated, this early partnership between filmmaker and government also included a corporate partner, the CPR. Each party benefitted from Freer's UK film tours. Freer was encouraged by the Canadian government and the CPR to make more films, and the government even sponsored a return tour of England. The government achieved its goal of increasing immigration to Canada. The CPR also achieved its goal of increasing patronage as evidenced by its commissions of other filmmakers to shoot films of Canada for further promotional campaigns in the years that followed (Morris, 32).

Before the governments of Canada became committed to the film industry, the documentary thrived with a partnership with the railway industry. First appearing as a "star" of the early films of Lumière brothers and Freer, trains and their companies soon began to take on roles off-camera in Canada as mobile studios and executive producers, respectively.

2.5 Europe, 1898–1901

Around the same time in Europe, other experiments with non-fiction film were proving to yield similar results in affecting social change. The lessons filmed of medical practices in 1898 in Romania and in 1899 in Paris were used to extend the reach to students, much like online courses do today. They were also used to demonstrate surgeries in close detail to students who could not otherwise witness the procedure in conventional operation theatres. This early precedent of using documentary film in this manner is found in the work of Eugène-Louis Doyen (France, 1899) and Gheorghe Marinescu (Romania, 1898 to 1901). Both documented medical procedures for the scientific community. One of Marinescu's most widely distributed films was called *Walking Troubles of Organic Hemiplegy* (1898). These medical instructional films were not intended to affect policy at the time, nor were they intended for the public at large, but they do represent the first attempts to document scientific data on film for the purpose of education. The success of these early cinematic experiments led to an increase in educational reach for the medical community. It might be argued that as an indirect consequence, these films resulted in a positive social change by enhancing educational practice for surgeons. Marinescu wrote that the role of film is "to complement and even replace … the descriptive exposition of phenomena by more rigorous, more exact analysis, which consists of (film) recording" (Barboi et al. 2004). As we will see in subsequent chapters, the factual—and perceived—truthfulness of the medium of both non-fiction film and science made for a powerful, if not obvious, union, a union that both the public and government policymakers trusted in providing visible evidence for the moulding of social change.

In England, an American entrepreneur named Charles Urban was building a studio business for non-fiction film by producing and selling films that followed the lead of Marinescu and Doyen in capturing medical procedures and other scientific data such as microscopic images in the educational and archival service to science. He also saw the power and the value of the cinematic technology as a communications tool for the state, specifically promoting its use for military applications (Urban, 3–5).

2.6 Canada, 1902

Corporate interest in documentary filmmaking was growing very quickly in Canada. After sponsoring Freer's original films and tours, like the Canadian public, the CPR wanted more. Freer, now a father of eight, abandoned film production after a less than successful second tour of England, so the CPR turned to Urban to make films that would promote settlement in the "Last Best West." The railway's involvement proved to be a key element in the early days of Canadian film production (Morris, 33). The film series, known as *Living Canada*, featured work by one of the most prominent camera operators of the day, Joe Rosenthal, perhaps best known for capturing the first battlefield footage when he filmed the Boer War in 1899. Dispatched by Urban to lead a team of camera operators through Canada, Rosenthal's CPR film series showcased the expansive breadth of the nation to those already living in Canada as well as to those considering moving there. The CPR also served as a film studio on wheels, providing funding, infrastructure, transport and even equipment to Canada's growing community of filmmakers.

So successful was this new method of promotion for the railway that it supported further advertising campaigns overseas through colourful print advertisements in newspapers, magazines and posters (see Fig. 2.2). Building a transcontinental railway was an expensive enterprise for the CPR, so the railway executives had to explore every option of advertising and promotion to generate revenue. To this end, they developed strategic partnerships and employed every promotional tool available to them at the time, including film, according to film historian Greg Eamon:

> The CPR in cooperation with the federal and provincial governments and with the Hudson's Bay Company, developed plans to encourage immigration and settlement to western Canada and the development of agriculture, mining and forestry. In order to meet these objectives the CPR developed an extensive system of promotion which included the use of still photographs, illustrated lectures (films) and testimonial pamphlets. (Eamon, 14)

Not to be outdone, other railway companies soon got into the act. The Canadian Northern Railway Company (CNRC) hired Urban under its own contract in 1909, to make films that would "induce settlers of *the*

Fig. 2.2 "The Golden Northwest, Home for All People, The Canadian Pacific Railway" (1902), Los Angeles: *AllPosters.com.* Reference Number: 11966290

right kind to emigrate from (Britain)" (Morris, 37). Also jumping on the bandwagon was yet another railway, the Grand Trunk Railway Company, who hired British filmmakers Frank Butcher and E.L. Lauste to travel their railroads and make films that would showcase the "fine specimens of the Anglo-Saxon race" (*The Bioscope*, 21) already settled in Canada in the hope of luring more.

Similar to Urban's films for the CNRC, the films made by Butcher and Lauste profiled such traditional Canadian scenes as Niagara Falls, harvesting, lumbering, native communities, and, naturally, the laying of new railway tracks. When Clifford Sifton financed the films of James Freer and his UK tour "under the auspices of the Canadian government" (Morris,

32), he sent a message that the new communications tool—film—was an effective influencer of the masses. As a result, the CPR and other train companies eagerly participated as producers of new promotional films to attract even more new customers domestically as well as internationally.

When the CPR expanded its film production department, this established the documentary's use by the corporate sector as a commercial promotional tool, equally influential as a communications tool reaching large groups of people, but more with a goal of profit than of a positive social change. Indirectly, these films did, in fact, result in positive social change for Canada by providing the revenue to unite the country by rail. Their success inspired other railway companies to use the same technology in their promotional campaigns.

2.7 England, 1907

In England, following the success of his *Living Canada* film series, Charles Urban began stockpiling films of a scientific, educative, and militaristic nature to sell to schools, universities, medical colleges, and governments. He also became one of the first theorists in non-fiction film when in 1907, his company, the Charles Urban Trading Company, published one of the first books detailing the process of filming non-fiction for educational and state use: *The Cinematograph in Science, Education and Matters of State.*

Urban's book details extensively the various sciences that could benefit from filmed recordings, commonly called the "cinematograph" at the time, and how they can be used in the classroom to enhance the educational experience of the student. The book also includes a list of scientific subjects that the Charles Urban Trading Company had available on film for rent to schools, many of these subjects microscopic in nature.

Another section of the book documents one of the earliest publications steering the use of film into the hands of government. The "matters of state" the book details are primarily military. The chapters in this section of the book include "The Cinematograph in Naval and Military Use," The Launching of War Vessels," "Maneuvers and Tactics," "Military Operations," and "The Cinematograph as a Recruiting Agent" (Urban, 3–5).

Throughout the book, Urban makes several claims to the medium's ability to reveal the truth as its leading virtue: film, Urban states, can present "a truthful and permanent record" (Urban, 46) as well as an "automatic and unerring record" (Urban, 56). Showing a disdain for those using film for fictional storytelling purposes—"The entertainer has hitherto monopolized for exhibition purposes, but movement in more serious directions has become imperative" (Urban, 7)—similar to Dziga Vertov's *kinopravda*'s (film truth) take on the same subject—"The film drama is the opium of the people ... down with bourgeois fairy-tale scenarios ... long live life as it is!" (Vertov, 71)—Urban's idea of the documentary film is far from Grierson's "*creative* treatment of actuality" (Grierson 1933), yet, at the same time, Urban promotes its use as an instrument of social improvement, a goal shared by Grierson. Urban concludes his book with this:

> [T]he Cinematograph has become, not—as some people imagine it to be—a showman's plaything, but a vital necessity for every barracks, ship, college, school, institute, hospital, laboratory, academy and museum; for every traveller, explorer and missionary. In every department of State, science and education, in fact, animated photography is of the greatest importance, and one of the chief and coming means of imparting knowledge. (Urban, 52)

Urban's book formalizes film's use in the scientific community, referencing the recording of surgical operations for training new doctors that was pioneered by Marinescu and Doyen. It also addresses the medium's ability to teach by providing films on microscopic images for educational purposes. Furthermore, the book details the use of film for government purposes—specifically military applications—representing one of the first texts identifying the state as a user of film to influence those it governs.

But perhaps one of the book's least known, yet most lasting, contributions to the history of film is the first English-language use of the word "documentary" to describe these kinds of educational and informational projects. The word *documentary* has long been used to describe a genre of film production characterized by non-fiction narratives. Recently, it has been undergoing some renovation as the documentary film is remediated in the digital domain: the database documentary, the interactive documentary (i-docs), the multilinear documentary, web-docs, eco-docs, among others; however, the first use of the word in English did not identify a type of film as a noun, but instead described a style of film as an adjective. Its use in this regard—as well as its first appearance in film terms—has been widely attributed to documentary pioneer John Grierson when he first used it in a review of Robert Flaherty's ethnographic docu-dramatic film account of a family in Samoa called *Moana* (1926). Grierson's review of this film in the *New York Sun* on February 8, 1926, included, in part, this now-famous description: "Of course, *Moana*, being a visual account of events in the daily life of a Polynesian youth and his family, has documentary value" (The Moviegoer 1926). Since then, this review, written by Grierson under the pen name "The Moviegoer," has been pointed to by many to be the first time the word "documentary" was used in film terms in English, often crediting Grierson with coining the term (Winston 1988; Renov 1993; Aufderheide 2007; McLane 2012; Rose 2014; Druick and Williams 2014; Nichols 2017, among many others).

While many scholars are convinced this marks the first appearance of the word documentary in English and in film terms, there have been some predecessors. Much earlier, similar uses of the word in French—"documentaire"—were made by Polish writer and filmmaker Bolesław Matuszewski in his French-language essay: "Une nouvelle source de l'histoire," published in Paris, on March 28, 1898. In describing the benefits of film as an historical artefact to display in museums, he also uses the word as an adjective:

> Forcément restreinte pour commencer, cette collection prendrait une extension de plus en plus grande à mer sure que la curiosité des photographes cinématographiques se porterait des scènes simplement récréatives ou fantaisistes vers les actions et les spectacles d'un intérêt *documentaire*. (*Italics are mine*) (Matuszewski, 7)

This translates as:

> This collection, which is necessarily restricted to begin with, would take on an increasingly large scale at sea, to be sure that the curiosity of film photographers (alt. cinematographers) would (result in work that) range(s) from merely recreational or fanciful scenes to actions and performances of *documentary* interest. (*Parenthetical additions and italics are mine*)

Both Grierson's and Matuszewski's use of the word documentary is descriptive—respectively, one of "value" and the other of "interest"—rather than as a noun as we commonly use it today (*I just watched a documentary*). It is used as a lawyer uses "documentary evidence" to differentiate from evidence that is unsubstantiated or mere hearsay; however, there is yet another documented use of the word in film terms, as an adjective, and in English, that pre-dates Grierson's "first use" by nineteen years.

The Cinematograph in Science, Education and Matters of State, written and published in 1907 by Charles Urban, a contemporary of both Grierson and Matuszewski who was also working in non-fiction filmmaking, the word "documentary" appears in the first sentence of the second paragraph in a chapter titled *A Cinematographic Course on Operative Surgery*, found on page 33 (see Fig. 2.3): "The Cinematograph will also allow of the

A Cinematographic Course on Operative Surgery.

"It would appear from this that the Cinematograph would be of great value in the course on Operative Surgery that all students should attend before entering the operating theatre. In this way, overcrowding of the amphitheatre would be avoided, and they would no longer hinder the surgeon without improving themselves, as is at present the case, since the majority see nothing, and those that do see have not sufficient knowledge to understand.

"The Cinematograph will also allow of the preservation in documentary form of the operations of the older surgeons. How valuable it would be to see again to-day upon the screen the operations of Langenbeck the elder, of Maison-Neuve, of Volkmann, of Billroth, or of Péan. The documents that we shall have henceforth will, thanks to the Cinematograph, allow the surgeon of the future to judge better of the progress achieved.

33

Fig. 2.3 "A Cinematographic Course on Operative Surgery" (1907), *The Cinematograph in Science, Education, and Matters of State*. London: The Charles Urban Trading Company, April, 1907. Page 33. http://www.cineressources.net/consultationPdf/web/o000/332.pdf. (Source: French Cinematheque-WD Film Library 8 ° 178 Br)

preservation in *documentary* form of the operations of the older surgeons" (*italics are mine*).

This early book of film theory, with its proposed methods for using film as an educational tool for science, as a training tool for the military and as a promotional tool for business, may be one of the first how-to guides for governments, educators, and corporations to use the relatively new medium of film for purposes more socially influential than mere entertainment. As significant as this may be historically, Urban's book may best be remembered more for its early use of the word *documentary* in English, nearly twenty years before John Grierson was credited with its invention.

2.8 Canada, 1919

The Canadian railway companies were not the only corporate entities to discover the promotional power of the documentary film. To commemorate the 250th anniversary of the Hudson's Bay Company (HBC), a film was commissioned to showcase HBC's work in Canada's north. The film was called *The Romance of the Far Fur Country* (see Fig. 2.4) and was

Fig. 2.4 "The Romance of the Far Fur Country" (1919), Winnipeg: *Hudson's Bay Company Archives*

made in 1919 by the New York company Education Films ("About the 1920 Film"). It was a two-hour feature documentary film, pre-dating what many consider to be the "first" documentary feature, Robert Flaherty's *Nanook of the North*, by three years. In fact, the narrative and filmic techniques used in HBC's film were later employed by Flaherty in making *Nanook* ("The Romance of the Far Fur Country"). Coincidentally, both films were made in and about life in the Canadian Arctic. *The Romance of the Far Fur Country* was directed by American cinematographer Harold M. Wyckoff (Geller, 244) who was sent by the HBC, together with company representatives Edmund Mack and Thomas O'Kelly, to ensure the film "projected an image of HBC's northern involvement from a company perspective" (Geller, 96). The HBC cleverly incorporated the beauty of the Canadian Arctic, the people and culture of the North and a "touch of romance" (Geller, 96) so the film would appeal to a general audience, not just already established customers and stockholders. This demonstrates an early use of the documentary film by the corporate sector as propaganda to promote its own profit motive. It also demonstrates the power of the documentary film to influence its audiences, a necessary function for effecting social change.

Robert Flaherty is often credited with making the first feature-length documentary film, *Nanook of the North*, focusing on an ethnographic look at an Inuit family in the Canadian Arctic in 1922. We now know *The Romance of the Far Fur Country* came earlier and told a very similar story; but since it was made as a promotional tool and commercial entity for the HBC, it did not receive the kind of wide theatrical release that *Nanook* enjoyed. As a result, *Nanook* became the benchmark of this new genre of filmmaking that presented unscripted non-fiction to audiences of places and people they would otherwise not know. It later came under fire for staging, but it still set the standard for the feature-length version of the genre.

Flaherty might have been heralded as the "father of documentary film" (Knopf, 209) by film scholars Kerstin Knopf, Richard Griffiths and others, but Canadian film historian Peter Geller reminds us that "what has been forgotten is that the HBC film shot in 1919 used many of the filmic and narrative techniques to tell its 'Life Story of the Eskimo' that Flaherty would later employ in his film. And outdoing Flaherty, the HBC film used titles in the Inuit language" (Nikkel 2012). Despite its profit underpinnings the film promotes, however unintentional, an authentic cultural component is also present, providing an archival artefact of Canadian native communities at the time.

These examples illustrate a common goal of using the early documentary film as a promotional tool by both the Canadian government (for immigration recruitment) and the corporate sector (for customer recruitment). Both organizations commissioned the filmmaker and gave specific instructions on what to shoot to achieve their respective promotional objectives. The artistic contribution of the filmmaker was compromised as Freer and his early filmmaking counterparts were eventually manipulated by both government and corporate employers, rendering the early Canadian documentaries more as products of function than as works of art. Thus, in its early days, the Canadian documentary film was used primarily for state and corporate agendas. Its effectiveness in this regard would soon inspire the Canadian filmmaker to use this powerful communications medium to trigger social change by presenting social issue films to the same government that pioneered this process of propaganda.

This alliance between the Canadian government and the railway companies continued to flourish and expand. The competition between the railway companies in Canada for the best film crews and the most film production became Canada's version of the Hollywood studio system. Having worked closely with the Canadian government in building the railways, and with government officials seeing the strength of the medium in stimulating immigration, a natural partnership coalesced.

The railway companies' sideline business in film took a major turn when the CPR established its own studio in 1920. The Associated Screen News of Canada (ASN) was incorporated by the Canadian Pacific Railway in Montreal. Ben Norrish, formerly of the Canadian Government Motion Picture Bureau (CMPB), was appointed as the head of ASN. From 1920 to 1958, ASN produced the majority of newsreels, shorts, and industrial films in Canada (Wise, 231). The significance of this reveals a closer collaboration commercially with the government and the corporate sector. As the CPR was now in the business of making movies for entertainment of the masses as well as corporate videos for other commercial clients, the appointment of Norrish kept the door open to government involvement with the content of newsreels.

The new emphasis on CPR's films indicates a shift in public demand. The ASN films were focused more on current events rather than on beauty shots of Canada's natural landscape, demonstrating an early desire to have films serve journalistic and educational purposes, more than self-promotional ones. The documentaries pioneered by the railway companies of Canada were mainly shorts, but soon gave way to a longer format of the genre. This early partnership between art and commerce provided a business model for a more formalized relationship between the Canadian provincial and federal governments and their respective filmmakers.

2.9 The Emergence of Government Participation in Documentary Filmmaking

This section examines the formal structures of government-run film studios in Canada on provincial and federal levels.

2.9.1 Canadian Overview

Federal Minister of the Interior Clifford Sifton launched the government's involvement into an emerging film industry in Canada in 1901 when he decided to back a second tour of Freer's films in England under the auspices of the Canadian government. The modest success of this first tour of Freer's films established a clear precedent of film's ability to influence the masses in Sifton's mind, prompting him to adopt this new technology, and not only fund its continued use for promoting immigration but also to brand the "Canadian Government" as the producer, showing the world how technologically advanced they were. At the time, Lord Sifton wrote:

> The CPR has initiated a series of animated photographs (films) of Canada, its scenery and its industries, which is much in demand. Naturally my department cooperates in any efforts that have for their object the dissemination of knowledge about Canada. (Eamon, 16)

This tour showcased many of the same films Freer first screened under the sponsorship of the CPR. Only one new film made by Freer was included: a trip across the Atlantic from Liverpool to Quebec City, presumably made upon his return to Canada from his first tour of England (Morris, 32).

Peter Morris believes that the Canadian government's involvement in documentary film production is "unique" and "among the most significant defining characteristics of film in Canada" (Morris, 127). The federal government's experiment with film was successful enough that a formal national film entity would be established about twenty years later, namely, the National Film Board of Canada (NFB). However, before we explore how this national film office was formed and used by the federal government, we need to examine its state-run predecessors to understand how they set the stage for the formation of the NFB since they established the methods and strategies that proved to be successful in reaching the public with targeted messaging.

2.9.2 *Government Film Offices in Canada, 1917 to 1924*

The new wave of immigration that peaked in 1917 introduced new challenges to the provincial and federal governments of Canada. Film bureaus began to become established provincially; however, the mandates of these offices were significantly different from the federal government's first use of film. Now that Lord Sifton had successfully used film to populate rural, Western Canada with immigrant farmers, a new challenge emerged: providing education for them. Since many of the new Canadians at the time were farmers with large families, they often lived in areas inaccessible to schools. The first provincial film office in Canada employed film to educate and to inform these remote people, making education its priority.

2.9.3 *The Ontario Motion Picture Bureau, 1917*

Government involvement in film production in Ontario dates back to 1914 when a Toronto production company, W. James and Sons, was hired to produce one of the first "making of" documentaries—the making of the Queen Elizabeth highway between Toronto and Hamilton (Morris, 137). At the same time, the Department of Agriculture commissioned educational films intended to train farmers.

In May 1917, the Ontario government established the Ontario Motion Picture Bureau (OMPB), making it the first state-run film organization in the world (Peter Morris et al.). At this time, a sociological shift in the use of the documentary was made when Ontario set up its film office. For the first time on a government level, the documentary was no longer going to be used to promote tourism and immigration, but rather, education and social improvements. The mandate of the OMPB was to provide "educational work for farmers, school children, factory workers and other classes" and to produce films that would encourage "the building of highways and other public works" (Peter Morris et al.).

One of the first filmmaking partners of the provincial film office was Pathescope of Canada, but not for the reasons you might think. The company was the Canadian agent for the French-based manufacturers of 28 mm film stock. What was unique about this particular brand of film was that it was not made of the highly flammable nitrate stock that 35 mm film stock was made of at the time. Its diacetate film composition was not flammable. Since most of the films produced by the OMPB were to be educational films shown in schools, there would be a substantial fire hazard using traditional 35 mm film stock, so a decision was made to use the "safety stock" of 28 mm exclusively available only through Pathescope of Canada. Naturally, they got the contract to produce as well. Despite this lucrative government contract, the manufacture of 28 mm film discontinued altogether only three years later, being replaced by 16 mm ("28 mm film").

Unlike the railway companies, the OMPB hired domestic filmmakers only, most of them from Toronto, establishing the government's focus of stimulating the home economy and growing the Canadian film industry from within. This employment priority was the first in a long line of commitment to Canadian content, both on and off the screen, and in many future provincial and federal government-run film offices. This approach was first identified by the OMPB "for the purpose of preserving Canadian traditions." In the years to come, we would see this mandate evolve into one that not only preserved Canadian traditions, but provided employment incentives to Canadian filmmakers in the form of public funding programmes, thereby strengthening the ties between government and filmmaker by becoming more of a corporate partner. Telefilm Canada, for example, not only provides government grants to filmmakers, but actively participates as a business partner with market support, investment programmes, and production loans.

The first director of the Ontario film bureau, S.C. Johnson, coordinated the production and distribution of films. This, too, was an original approach to making the domestic film industry sustainable. When the railway companies were producing films, they often hired British filmmakers who regularly came with foreign distribution plans. Since the OMPB was more interested in educating a domestic audience, foreign distribution of the Canadian documentary was not a priority.

About five years in, the OMPB bought the Adanac (Canada spelt backwards) Producing Company's Trenton studio to house its film bureau (Melnyk, 31). The OMPB not only had office space, but, unlike any other government office, it now had its own studio space, dedicated to producing films for the purpose of educating all classes and improving their lives. By 1925, the OMPB became a major player in the world's new film industry, distributing as many as 1500 reels of film every month (Melnyk, 42). But this boom was short-lived in Ontario. Technological advances, such as sound and a film stock format change from 28 mm to 16 mm, provided a strain on the resources of the OMPB, and in 1934, the bureau was shut down by the provincial government (Melnyk, 42).

This version of the early Canadian government documentary production system saw the corporate sector of the triumvirate of filmmaker-government-corporate take a back seat. There was no profit incentive in this approach to documentary filmmaking as the goal of the government in this case was education and using the filmmaker to create the products that would achieve this specific goal of positive social improvement.

2.9.4 *The British Columbia Patriotic and Educational Picture Service, 1919*

The provincial film office in British Columbia adopted a mandate to foster economic development for Canadian industry and trade in addition to education. The British Columbia Patriotic and Educational Picture Service (BCPEPS) was quite clear about the kinds of films it was to provide: "films … of a patriotic, instructive, educative, or entertaining nature; and, in particular, … films … depicting the natural, industrial, agricultural or commercial resources, wealth, activities, development, and possibilities of the Dominion" (Gasher, 32–33).

Particularly unique to the province's desire to produce films about its industries and potential economic possibilities to foreign and domestic interests was its mandated quotas to BC's movie theatres to show these films. Under the Department of the Attorney-General, a quota provision required BC movie theatres "to introduce each film program with fifteen minutes of films either produced by, or approved by, the Picture Service" (Gasher, 32).

While government involvement in documentary film production in Canada was more of a collaborative effort with filmmakers, the provincial government in BC was the first to start "calling the shots" on private industry. The quota represented "the first government film unit in North America with statutory authority to compel the screening of its productions" (Gasher, 33). This pioneering directive further eliminated the need to engage British filmmakers who had previously been able to guarantee exhibitions in England.

This represents a new shift in the relationship of the triumvirate of filmmaker-government-corporate sector. Instead of being a mere tool of the government, the Canadian documentary filmmaker was now being treated as a valuable employee, and the quota was designed to keep its filmmakers working. It also represents the first time the government imposed content restrictions, a practice that was severely criticized at the time (Gasher, 32). Corporate opposition was incessant, and it even became a political issue in the 1920 provincial election. As a result, the fifteen-minute quota was no longer enforced by 1924 (Gasher, 33).

2.9.5 *Quebec's Ministry of Agriculture and Education, 1920*

The next province to join the ranks of government film studios was Quebec. Instead of establishing its own agency, the provincial government conducted its film operations through its Ministry of Agriculture. The films produced here focused on education as their social improvement mandate. The film division operated in this manner for twenty-one years until Quebec established Le Service de Ciné-Photographie in 1941 to handle all fiction and non-fiction film production made by the province and its citizens (Elder, 98).

Launching the provincial government's foray into documentary film production, distribution and exhibition was Joseph Morin, a civil servant with the Ministry of Agriculture, who, "as early as 1920 had the idea of using film for pedagogical ends" ("Cinema in Quebec"). Morin carted around four documentary films to the farmers of rural Quebec, foreshadowing the habitual practice of the National Film Board of Canada years later. He often brought his own generator as well to ensure that power was available for operating his projector in areas not yet serviced with electricity ("Cinema in Quebec").

Using his films for the purpose of educational outreach, Morin found great success with this new medium, prompting the importation of other documentary films from the United States and France. His film archive was the first in Canada and served as a model for other government departments ("Cinema in Quebec"). This provided the government in Canada with a more powerful resource in its future propaganda campaigns by having access to a wide selection of international content to support its own messaging to the Canadian public.

Unlike the other provinces in Canada with film offices at the time, the Catholic Church in Quebec was not a fan of film. In 1916, the *L'Action Catholique* newspaper began an investigation into Quebec cinemas in an attempt to fight films they claimed promoted "debauchery and scandal" ("Cinema and Institutions"). Throughout the province, "priests and bishops declared cinema the cause of all evils and the main enemy of French-Canadian identity" ("Cinema and Institutions"). As a result, many of the educational films made in Quebec during these early years were made by

Catholic priests in order to ensure the preferred content of the Church (Véronneau, *The Cinema of Quebec*). One of the most prominent "Father Filmmakers" was Rev. Albert Tessier. From 1925 until his death in 1976, he made more than seventy documentary films, making him one of Canada's most prolific documentary filmmakers of all time ("Albert Tessier"). Sanctioned by the Church, many of his educational film subjects were about nature, Quebec history, religion, and how "strong women" (1938) were becoming "social misfits" (1948) ("Albert Tessier"). Once again, we see an example of documentary film being used as propaganda by an institution in Canada, this time the Catholic Church.

Having received his PhD in theology from Rome in 1922, Tessier moved back to Quebec and immediately began using the new medium of film to produce educational films primarily about agriculture and nature. His devotion to using film as an educational tool to "build and fortify" (MacKenzie, 11) rural communities resulted in the City of Paris honouring him with the French Academy Medal in 1959, recognizing his films for "their influence on the evolution and progress of the French life in Canada" (Groulx, "Tessier, Albert"). Twenty-one years later, the province of Quebec introduced the Albert Tessier Prize honouring "outstanding careers in Quebec cinema." The recognition, both at home and abroad, of the contribution to documentary film in Canada by Father Tessier pays tribute to his community-based filmmaking technique, a method that would later become known as the participatory mode of documentary filmmaking. He developed this technique to better represent the voice of those he profiled in order to improve their lives. Both Church and government responded to these films by providing aid to these rural, northern Quebec communities, demonstrating the success of this method.

According to Quebec cinema historian Scott MacKenzie, the much-heralded work of Father Tessier represented a shift in Quebec filmmaking, "from feature films to documentary, from entertainment to socially activist filmmaking" (MacKenzie, 10). MacKenzie explains that Father Tessier travelled the province taking his films to rural areas and screened them with ad-lib voice-over. Tessier would then re-edit the films taking into account the discussions he had had with his audiences. According to MacKenzie, this process had a great influence on the NFB's *Challenge for Change* (CFC) series.

> What is of interest here is … the concern with building a discursive community through the use of cinematic images. This desire pre-dates … the theoretical and practical strategies developed in the *Challenge for Change/Société nouvelle* programmes of the 1960s. (MacKenzie, 10–11)

Tessier's discursive community approach inadvertently represents one of the first attempts at participatory documentary filmmaking in Canada. In establishing a standard of community involvement in the documentary filmmaking process, and with the social issue slant of the CFC's use of this technique, films made this way are seen to represent a kind of "reverse propaganda." The messaging in these films came from the bottom-up, from the public to its government, instead of from the top-down, as was previously the successful model. Tessier's innovative method was mirrored by Colin Low in the CFC film series of the NFB. Appropriately, Low's success with Tessier's technique was later honoured by the province of Quebec when Low become a recipient of the Albert Tessier Award in 1997.

This participatory approach is one of the keystones of successful documentary filmmaking whose intentions are creating positive social change, a technique that will be examined more extensively in Chap. 3.

2.9.6 The Motion Picture Branch of Saskatchewan's Bureau of Publications, 1924

Following the lead of Ontario, Quebec and, to a lesser extent, British Columbia, Saskatchewan also created a governmental film division to produce educational films in these formative years for Canada's emerging documentary film industry. Its Bureau of Publications at the time was busy promoting the province as a tourist destination (see Fig. 2.5), so the time seemed right for the Bureau to open a branch for that most powerful promotional medium: the motion picture. The branch was headed by professional agriculturist Harry Saville. As one might expect, many of these films dealt with issues related to prairie farming: *Hog Raising in Saskatchewan*, *Saskatchewan's War on the Grasshopper*, and *Farm Boys' Camp, Regina* (Morris, 151).

At the time, the majority of Saskatchewan film audiences were in some way involved with farming. As a result, many of the subjects of these documentaries were related to farming. With this as a priority, the Saskatchewan government film office concentrated on using film primarily for educational purposes on topics predominantly dealing with agriculture.

Collectively, these early forays into documentary film production by the state in Canada reveal an understanding and enthusiasm for the medium's ability to bring about social change. Specifically, their focus represented the social improvement areas of immigration, education, trade, and commerce. As we have seen, the collaboration between the government and

Fig. 2.5 "Beautiful Saskatchewan, Canada: The Playground of the West" (1924), Regina: *Saskatchewan Bureau of Publications*. Reference Number: Peel 9903

the corporate sector in using the filmmaker to realize their respective goals of growing the population and the national economy was effectively realized by these provincial film offices. These early experiments in government film production, exhibition, and distribution provided models of what worked and what failed when the National Film Board of Canada was formed and created its own mandate.

Canada's other provinces formed their own film offices in the years to come, but it was these first provincial government film departments, together with the federal version of them, the Canadian Government Motion Picture Bureau, that laid the groundwork for a new federal film organization that would quickly define Canada as a world leader in social change and activist documentary filmmaking: the National Film Board of Canada.

2.9.7 National Film Offices in Canada, 1918 to 1939

The federal counterpart of the OMPB opened its doors in 1918, one year after Ontario opened its doors. It, too, had a distinctive first to its credit: the first national state-run film organization in the world (Peter Morris et al.). Originally named the Exhibits and Publicity Bureau, it was later renamed the Canadian Government Motion Picture Bureau on April 1, 1923. The word "Government" was eventually dropped, and it was more commonly known as simply the CMPB or the "Bureau" (Melnyk, 41). It should be noted here that a national predecessor to the Bureau was founded in 1914, but it provided largely scenic and travel films, lacking the social relevance that many Canadian documentary films had featured (McLane, 131).

Unlike its provincial predecessor, the CMPB focused its lens on Canada's trade and industry. While Ontario was using the documentary film to educate children, rural farmers, and the working classes, the Canadian bureau was more interested in following the BC model in using film to stimulate the national economy by showcasing Canadian industry to foreign and domestic interests. As the Federal Minister of Trade and Commerce, Thomas Andrew Low, then said: "the Bureau was established for the purpose of advertising abroad Canada's scenic attractions, agricultural resources and industrial development" (Peter Morris et al). While using film to showcase Canada's natural splendour, as Clifford Sifton and the CPR did with the films of Freer and Urban to stimulate immigration to Canada, the CMPB was using these "beauty shots" to stimulate business interests, both domestically and internationally.

Evidence of this is the body of documentary short films produced by the CMPB called *Heritage* (1939) and *Farmers of the Prairies* (1940). These films introduced new farming techniques to stimulate the agricultural industry in Canada and to assist its new farmer immigrants. For audiences outside of Canada, the films served to position Canada as a leader in agricultural techniques and technologies. Canadian film political scientist Malek Khouri argues that these films set the groundwork for establishing the Canadian government as a knowledgeable and reliable business partner.

> (The CMPB films) explore how the intervention of the government helps farmers deal with their problems (and) that there are major benefits to be gained from having the government involved in creating agricultural aid programs and in introducing new scientific research. (Khouri, 109)

With more than twenty years of film production under its belt, the bureau had wanted to expand and sharpen its particular brand of social service vision of helping Canadians through economic stimulation by promoting Canada's industry and trade (Peter Morris et al.). To this end, the Bureau extended an invitation to British documentarian John Grierson in 1930 to come to Canada and assess its operation.

2.9.8 The Controversial Vision and Influence of John Grierson

When John Grierson accepted the invitation of CMPB to review its organization, it seems their commercial aims of promoting tourism and trade were not as interesting to Grierson as was the genre's ability to affect social change, as the films of Freer and the OMPB had done earlier. Philosophically, Grierson believed that film could do much more and be used to manipulate the masses to yield societal improvement.

Grierson's own education reveals the source of his proclivity towards this aim. His sociological ideologies were established in his graduate work at the University of Chicago where he was first introduced to the study of social sciences. The university was the first to introduce the social sciences to academic studies in America when it formed its Department of Sociology and Anthropology in 1892. In fact, "Chicago's was the first sociology department anywhere in the world" (Druick, 49). On a Rockefeller scholarship, Grierson pursued his master's degree in sociology there in the mid-1920s, and the subject of his thesis was "immigration and its effects on the social problems of the United States" (Druick, 49).

One of Grierson's greatest influences at the University of Chicago was his supervisor, Charles E. Merriam, author of *American Political Ideas.* In it, Merriam argued for the need for civic education and social scientific study of the population "to assess loyalty and devise plans for building national sentiment" (Druick, 52). Grierson used these theories to convince Canada to use films for the purpose of public education as a means of producing Canadian unity among its social binaries (French/English, male/female, urban/rural, East/West, native/immigrant) and create a sense of distinction from the United States. Merriam's influence was so strong, a copy of his book was reportedly always within Grierson's reach "until the end of his life" (Druick, 51).

His other supervisor, Robert Park, was similarly influential with respect to the philosophies Grierson brought to the National Film Board of Canada. Park focused on "social groups" rather than on individuals in his study of social problems within society, particularly urban centres. He found that "all social problems turn out finally to be problems of group life" and concluded that culture "perpetuates social life through education" (Druick, 53).

Just as "the invasion of an alien species" can disrupt a community of plants, so too can foreign ideas and customs disrupt a social community, he argued. While this may sound somewhat racist today, the focus of Park's conclusions was based on environments of social differences, not on the essential qualities of the individual. Education, he therefore asserted, can be an effective control mechanism to develop healthy social cultures for all.

These social theories shaped Grierson's vision on how film can be used to reach the masses and, despite the socio-political-geographical-ethnic-economic differences of the Canadian people at the time, deliver messages to all. The goal would be to forge a single, unified, national, social culture.

This study served him well in helping the Canadian government develop a programme to assist in its governance of the people of Canada through film. Upon graduating from the University of Chicago in 1927, Grierson returned to England, intrigued with the idea of applying his socialist theories and Robert Flaherty's film techniques to the common people of Scotland. He first sold his idea of documentary storytelling to the Empire Marketing Board, to make his first film, *Drifters*, in 1929 (Winston, 37). This silent film featured the harsh life of fishermen in the North Sea.

At this time, he began giving lectures at the London Film Society, detailing his new visionary approach to the way the documentary film can be used as an educational tool to help shape social cultures. At one of these lectures, Ross McLean, a secretary to the Canadian high commissioner Vincent Massey, heard one of Grierson's lectures and was inspired to write a report to Massey, recommending a government film service modelled after Grierson's social reform theories (Druick and Williams, 107).

As a result, Grierson received an invitation from the Canadian government to evaluate the Canadian Government Motion Picture Bureau in 1930. Grierson's initial report of the Bureau was harsh and critical of the films it produced, claiming they misrepresented "the real Canada" by only showcasing beautiful scenery and "people on holidays" (Hardy, 94). He also did not make any friends at the Bureau when he suggested it be staffed with non-Canadians. One of his most controversial recommendations was that "the best associate producers available in Great Britain or the United States should be sought to begin with, preferably from Great Britain"

(Nelson, 60). This created an irreparable antagonism between Grierson and the CMPB's director at the time, Frank Badgley (Nelson, 61).

According to film scholar Joyce Nelson, "Grierson's primary interest was in stark contrast" (Nelson, 61) to the types of films being made by the CMPB at the time. In Grierson's report, he recommended "a central organization" that would coordinate the mutual goals of Canadian nationalism and British imperialism (Druick and Williams, 108).

Grierson made many public and official assessments of the films of the CMPB. This representative sample is typical of his opinion:

> I have seen many Canadian films and most of them were about National Parks and people on holidays. I didn't, so help me, believe that Canada could be just the big innocent, baby-hearted holiday haunt it pretended to be in its pictures. I thought maybe somebody did some work once in a while and that Canada's work might just conceivably have something to do with the real Canada. That in fact is my interest in the world. I would like to see more and more films about real people. (Qtd. in Hardy, 94)

With reports such as this, Grierson's ultimate recommendation was to establish "a central organization which would co-ordinate demands and through which the Canadian Government and the British and Dominion film interests could work" (Druick and Williams, 108). It was this recommendation that eventually led to the formation of the National Film Board of Canada by the Department of Trade and Commerce in 1939. After his assessment of the CMPB, Grierson would be invited back as the first commissioner of the NFB. In this capacity, Grierson articulated that the aim of the documentary is to "bring about positive social change" (qtd. in Rosenthal, 6).

2.9.9 The National Film Board of Canada, 1939

The infrastructure of the NFB in its formative years comprised "a dominion film commissioner as CEO, two ministers of the federal government, three senior civil servants and three private individuals with interests in film" (Druick and Williams, 108). The National Film Act of 1939 set the table for the NFB's original mandate "to produce and distribute and to promote the production and distribution of films designed to interpret Canada to Canadians and to other nations" ("Missions and Highlights"). It is interesting to note that no provincial entities were invited to form this federal body.

While the CMPB attempted to work within the new NFB, the rift between Badgley and Grierson, along with the conflicting visions of the two organizations as to film content, style, purpose, and audience, made it difficult for the two state-run film entities to co-exist. The Bureau was ultimately absorbed by the NFB three years later (Druick and Williams, 108).

Grierson used the social theories he learned at the University of Chicago to convince Canada to use films for the purpose of public education as a means of producing Canadian unity. Grierson might deserve credit for using the word documentary as a noun, but his most valuable contribution to mobilizing the documentary film as an instrument of social change is evident when he demonstrated the benefits of education in public governance to the Canadian government. While the theory was embraced, there were challenges in its practice, namely, reaching the geographically dispersed Canadian population.

He suggested bringing the films to the people by engaging church groups, business clubs, and women's organizations to act as exhibitors. NFB projectionists would travel great distances to remote areas throughout Canada with their projectors and films. This methodology was equally embraced by the Canadian people themselves who "formed some 250 voluntary film councils across Canada and, besides assuming the projectionists' duties, began operating 230 film libraries" (Evans, 7) of NFB films.

This social science approach to delivering governance to the people through the films of the NFB was supported and encouraged throughout Parliament at the time. One MP, Roy Theodore Graham, said: "(a) national film board can interpret government to the people and help them be fully informed of the details of different regulations and laws that affect them in their daily lives" (Druick, 57). As Canada would soon see, the reciprocal relationship between its citizens and their government with film would prove to be equally effective.

Some scholars, like Gary Evans, attribute much of the NFB's early success to Grierson's determined and focused vision:

> He lived life as a Modern-day Prophet of Cinema, a visionary who founded the documentary-film movement, a teacher whose primary interest in film was its potential to act as an agent of social change. (Evans, 3)

Grierson's vision was proving to be effective. Canada's involvement in the formation and development of the documentary film in these early days had made it well-known and respected throughout the world as a leader in the production of this genre. In fact, the NFB would receive its first Academy Award for documentary film in 1941 for *Churchill's Island*, an examination of the strategy behind the Battle of Britain in 1940. The NFB was also the first recipient in this new category created by the Academy.

At this time, Grierson's theories were being put to their biggest test of all when the NFB entered the war effort. According to film historian Gary Evans, "Grierson and his associates used truth as their standard and propaganda as education in an effort to define public duty" (Evans, 3) during World War II. One of the most successful film projects was *The War for Men's Minds* (1943). The film was able to show Canadians the importance of their involvement in combatting enemy forces not just on the ground, but in their minds as well. Here is a passage from the film's narrative that not only defines its aims, but also showcases Grierson's vision of social reform through education in film:

> We are faced with a powerful and ruthless campaign to engulf the world with a philosophy and social system which is alien and repugnant to us. We are living in constant fear that this battle of words and ideas will turn into armed aggression and we are arming ourselves and seeking our allies to meet the threat of this aggression. (Evans, 20)

The Department of External Affairs encouraged the NFB to create a series of such films called *Freedom Speaks* (Evans, 21). Other war-effort film series at the time included *Canada Carries On* and *The World in Action*. After the shorts in these series were produced, a compilation film emerged in 1953, called *Germany: Key to Europe*. The twenty-minute film ignored how West Germany planned to deal with its legacy of Nazism, and consequently, the film was seen as a failure in the field of propaganda. Evans asserts that "the film stands as an example of why the Film Board was not a particularly effective cold warrior" (Evans, 21). While the NFB had established a particular expertise in propaganda filmmaking, it was less effective producing wartime propaganda films.

Grierson's insistence on the NFB having "final cut" on all films, including those made to support the war effort, eventually led to his resignation, as certain NFB films were seen as "Communist" in theme and messaging:

> *Our Northern Neighbour*, which portrayed a Russian soldier in the Allied forces in a favourable light, attracted heavy criticism. United Artists refused to distribute it in the U.S.A., and the Quebec Censorship Office blocked it too. The NFB was accused of creating Communist propaganda at the expense of the Canadian government. In January 1945, Stuart Legg's

> *Balkan Powder Keg* threw negative light on the British Government's actions in Greece and the prime minister ordered it withdrawn. This incident raised the question of the freedom of action of the NFB, but Grierson reaffirmed its independence and refused to give the Minister of Foreign Affairs any veto powers over films. The film was taken out of circulation, re-cut and re-baptized *Spotlight on the Balkans.* This faux pas put an end to the freedom that Grierson had enjoyed until this point, and he was not encouraged to remain at the head of the NFB after the war. Repudiated by the Canadian government and filled with ambitious projects for the international development of the documentary, Grierson resigned on August 7, 1945 and left the NFB definitively on November 7, 1945. ("The 1940s: The NFB")

This conflict between artist and government reflects the ruling power of the hegemonic state. Working even at arm's length, the Canadian documentary film's final representation and message lie with the state.

Even though he was seen as having a political agenda in these later years, Grierson's strategy was non-partisan. His intent with the NFB was "to use the apparatus and money of the liberal state to create universal humanitarian loyalties in the hearts of its citizenry" (Evans, 3). Applying this strategy and mission to the formation of the National Film Board gave rise to not only a unique government-sponsored film studio, but also a new purpose to the aim and goals of non-fiction filmmaking. Political scientist Patricia M. Goff describes the government in this relationship as one of "patron" and "catalyst" to the filmmaker. In her book, *Limits to Liberalization: Local Culture in a Global Marketplace*, she explains that "these guises are guided by an overall desire (of the Canadian government) to participate in cultural industries *at arm's length* so as to ensure freedom of expression (of the filmmaker)" (Goff, 56).

2.9.10 Propaganda Documentary Filmmaking in the United States

South of the border, the United States began to experiment with the state use of documentary film to promote its own propaganda messages to its citizens. The move was not embraced as much as it was in Canada, with the US film industry objecting to the government getting involved with the film business. One of the first government film contractors was New York film critic Pare Lorentz.

Following his graduation from Yale University in 1928, Lorentz almost immediately entered the professional world of film criticism, writing for the *New Yorker*, the *New York Evening Journal*, *Vanity Fair*, *Town & Country*, and *McCall's Magazine* (Roberts, 37). It was during his time in New York that Lorentz was greatly impressed by an unforgettable environmental catastrophe that took place one day at Times Square:

> [O]ne day in New York when I was working at Newsweek and a heavy, slow-moving, gray cloud, dust from the drought-stricken Great Plains, blew down in the middle of Manhattan Island and settled like an old blanket over the tower of the New York Times building in Times Square … (This) experience(s) made a lasting impression and led me to recommend the Dust Bowl … as the first subject … (of my) first movie. (Lorentz, 37–38)

His recommendation was made to Rexford Guy Tugwell, administrator of a new federal government organization at the time called the Resettlement Administration, to make a documentary film about the Dust Bowl that would be called *The Plow That Broke the Plains*. Through a mutual connection in the Office of Information—its chief, John Franklin Carter who wrote with Lorentz at *Vanity Fair*—Lorentz was introduced to Tugwell, who also held the title of undersecretary of agriculture at the time.

> Tugwell was so enthusiastic that he suggested that we make eighteen movies. I never did figure out why the number eighteen. Anyway, I thought we'd better make one first and see how it went before we scheduled more. He agreed. (Lorentz, 37)

While Lorentz made a very good documentary in his filmmaking debut, there seems to have been some naivety on his part and his government-producer as to exactly what kind of documentary they were making. Neither party saw what they were making to be propaganda, but that is exactly how their industrial rival, Hollywood, saw it. Hollywood hated Lorentz and his kind of filmmaking, refusing to give him any stock footage when he formally requested it. In desperation, he enlisted friends to sneak him into projection rooms and steal the footage he selected.

Furthermore, *The Plow* did not get the wide theatrical release it had expected. All the major eight distributors refused to release it, claiming its length (twenty-nine minutes) as the problem: too long for a newsreel; too short for a feature. But many claimed they did not want to show a government film. One distributor said, “I wouldn’t release any government picture, not even if it were *Ben Hur*!” And yet another explained that “if any private company or individual made this picture, it would be a documentary film. When this government makes it, it automatically becomes a propaganda picture” (Snyder, 43–44). The Academy Awards followed suit and banned it from competing in its annual awards.

The concern over state use of the documentary film as a propaganda tool was growing not only in North America, but across Europe as well. With Cold War tensions growing, Henri Langlois of the Cinémathèque Française in Brussels established the World Union of Documentary Filmmakers, an all-inclusive, international organization for documentary filmmakers. Its lack of exclusivity became a major issue with Grierson who did not believe certain Eastern European filmmakers should belong. Underwater documentary pioneer Jean Painlevé and poetic documentary artist Joris Ivens strongly disagreed with Grierson, and, ultimately, the organization could not overcome this division and disbanded the following year (Berg, 37). It did manage, however, to create a definition of documentary film current to its time at a conference in Czechoslovakia in 1948:

> Any film that documents real phenomena or their honest and justified reconstruction in order to consciously increase human knowledge through rational and emotional means and to expose problems and offer solutions from an economic, social, or cultural point of view. (Berg, 39)

The significant section of this definition is its acknowledgement that the documentary film has an obligation to promote social reform and its recognition that it has the ability to do this effectively.

2.10 Conclusion

To summarize, the invention of the motion picture quickly developed into a non-fiction chronicle of reality that science embraced for the purpose of education. Other subjects were soon explored, using this new educational tool primarily in schools and universities. The corporate sector soon made great strides in marketing and promotion using film, but the widest application of this new technology for social change emerged when the state began to use it to deliver its messaging to large groups of people, both domestically and internationally. Immigration recruitment, military training, ethnographic archiving, economic improvement programmes, communication, education, medical training, serving rural communities, and propaganda are just some of the early government uses of film when it became defined as documentary.

The second half of the twentieth century saw the refinement of motion picture technology and documentary practice focusing on theoretical applications aimed to enhance its natural ability to inform and influence. Along with these experimental practices came technological advances that better accommodated these theories and helped ameliorate the documentary film as an instrument of social change.

Bibliography

"28 mm film." Wikipedia, The Free Encyclopedia. July 14, 2016. Web. September 4, 2016. https://en.wikipedia.org/wiki/28_mm_film.

"A Daguerreotype of Daguerre." National Museum of American History, Smithsonian Institution. Web. Accessed December 12, 2017. https://ipfs.io/ipfs/QmXoypizjW3WknFiJnKLwHCnL72vedxjQkDDP1mXWo6uco/wiki/Daguerreotype.html.

"About the 1920 Film." *The Romance of the Far Fur Country*. Winnipeg: Five Door Films, n.d. Web. Accessed May 28, 2015. http://www.returnfarfurcountry.ca/romance_about_the_film.html.

"Albert Tessier." *Wikipedia, The Free Encyclopedia*. Web. August 14, 2014. Accessed June 12, 2015. https://en.wikipedia.org/wiki/Albert_Tessier.

Aufderheide, Patricia. *Documentary Film: A Very Short Introduction*. New York: Oxford University Press, 2007. Print.

Barboi A.C., C. Goetz, and R. Musetoiu. "The Origins of Scientific Cinematography and Early Medical Applications." *Neurology*. Milwaukee: Medical College of Wisconsin, Froedtert Memorial Lutheran Hospital, 2004. Web. Accessed March 22, 2015. https://en.wikipedia.org/wiki/Cinematography_in_healthcare.

Bazin, André. "The Ontology of the Photographic Image." Translated by Hugh Gray. Berkeley, University of California Press: *Film Quarterly* 13, no. 4 (Summer, 1960). Print.

Benjamin, Walter. "Paris, Capital of the Nineteenth Century." In *The Arcades Project*, ed. Rolf Tiedemann. Cambridge, MA [etc.]: Belknap Press, 1999. Print.

Berg, Brigitte. "Contradictory Forces: Jean Painlevé, 1902–1989." In *Science Is Fiction: The Films of Jean Painlevé*, ed. Andy Masaki Bellows, Marina

MacDougall and Brigitte Berg and trans. Jeanine Herman. Cambridge, MA: MIT Press, 2000. Print.

"Cinema and Institutions." *Silent Cinema in Quebec: 1896–1930*. Montreal: Cinemamuetquqebec.ca. Web. Accessed June 23, 2015. http://www.cinemamuetquebec.ca/content/theme_cap/57?lang=en.

Cinema in Quebec: The Talkies and Beyond. www.cinemaparlantquebec.ca. Web. Accessed August 17, 2015. http://www.cinemaparlantquebec.ca/Cinema1930-52/pages/textbio/Textbio.jsp?textBioId=36&lang=en.

Clegg, Brian. *The Man Who Stopped Time*. Washington: Joseph Henry Press, 2007. Print.

Druick, Zoe. *Projecting Canada: Government Policy and Documentary Film at the National Film Board*. Montreal and Kingston: McGill-Queen's University Press, 2007. Print.

Druick, Zoe, and Deana Williams, eds. *The Grierson Effect: Tracing Documentary's International Movement*. London: British Film Institute, 2014. Print.

Eamon, Greg. "Farmers, Phantoms and Princes. The Canadian Pacific Railway and Filmmaking from 1899–1919." *Cinémas: Journal of Film Studies* 6, no. 1 (1995). Montreal: University of Montreal. Print.

Elder, R. Bruce. *Image and Identity: Reflections on Canadian Film and Culture*. Waterloo: Wilfrid Laurier University Press, 1989. Print.

Evans, Gary. *In the National Interest: A Chronicle of the National Film Board of Canada from 1949 to 1989*. Toronto: University of Toronto Press, 2001. Print.

Frank, David. "In Search of the Canadian Labour Film." In *Working on Screen: Representations of the Working Class in Canadian Cinema*, ed. Malek Khouri and Darrell Varga. Toronto: University of Toronto Press, 2006. Print.

Gagnon, Erica. "Settling the West: Immigration to the Prairies from 1867 to 1914." Canadian Museum of Immigration. Halifax: 2015. Web. Accessed June 10, 2015. http://www.pier21.ca/research/immigration-history/settling-the-west-immigration-to-the-prairies-from-1867-to-1914.

Gasher, Mike. "Cinema as a Medium of Regional Industrial Development." In *Hollywood North: The Feature Film Industry in British Columbia*. Vancouver: UBC Press, 2002. Print.

Geller, Peter. *Northern Exposures: Photographing and Filming the Canadian North, 1920–45*. Vancouver: UBC Press, 2004. Print.

"Gheorghe Marinescu." *Monoskop*. Bergen, Norway: Bergen Center for Electronic Arts, December 11, 2013. Web. Accessed April 12, 2015. https://monoskop.org/Gheorghe_Marinescu.

Goff, Patricia M. *Limits to Liberalization: Local Culture in a Global Marketplace*. Ithaca, NY: Cornell University Press, 2007. Print.

Gray, James H. *Red Lights on the Prairies*. Toronto: Signet Books, 1973. Print.

Grierson, John. *Cinema Quarterly* 2.1, Edinburgh: G.D. Robinson, 1933. Print.

Groulx, Lionel. "Tessier, Albert." *Encyclopédie de L'Agora: Pour un Monde Durable*. Quebec City: L'Encyclopédia de l'Agora, 2012. Web. Accessed June 11, 2015. http://agora.qc.ca/dossiers/Albert_Tessier.

Hardy, Forsyth. *John Grierson: A Documentary Biography*. London: Faber & Faber, 1979. Print.
"History of Film Technology." *Wikipedia, The Free Encyclopedia*. Web. Accessed May 11, 2015. http://en.wikipedia.org/w/index.php?title=History_of_film_technology&oldid=661858242.
"Immigrant Voice." *CanadianHistory.ca*. Web. Accessed August 16, 2015. http://www.canadianhistory.ca/iv/1867-1914/laurier_boom/sifton4.html.
Khouri, Malek. *Filming Politics: Communism and the Portrayal of the Working Class at the National Film Board of Canada, 1939–1946*. Calgary: University of Calgary Press, 2007. Print.
Knopf, Kerstin. "Electronic Storytelling in the Arctic." In *Transcultural English Studies: Theories, Fictions, Realities*. Amsterdam and New York: Editions Rodopi B.V., 2009. Print.
Les films Lumière. Lyon: L'Institut Lumière, 2015. Web. Accessed June 14, 2016. http://www.institut-lumiere.org/musee/films-lumiere.html.
Lipovetsky, Gilles. "A Century of Fashion." In *Fashion Theory: A Reader*, ed. Malcolm Barnard. London; New York: Routledge, 2007. Print.
Loiperdinger, Martin, and Bernd Elzer. "Lumiere's Arrival of the Train: Cinema's Founding Myth." *The Moving Image* 4, no. 1 (2004). Print.
Lorentz, Pare. *FDR's Moviemaker: Memoirs and Scripts*. Reno, Las Vegas, London: University of Nevada Press, 1992. Print.
Macdonald, Kevin, and Mark Cousins. *Imagining Reality*. London: Faber and Faber Ltd., 2006. Print.
MacKenzie, Scott. *Screening Québec: Québécois Moving Images, National Identity, and the Public Sphere*. New York: Manchester University Press, 2004. Print.
Matsuda, Matt K. *The Memory of the Modern*. New York: Oxford University Press, 1996. Print.
Matuszewski, Bolesław. *Une nouvelle source de l'histoire*. Paris: March 28, 1898. Print.
McLane, Betsy A. *A New History of Documentary Film* (2nd ed.). New York and London: Bloomsbury Publishing, Inc., 2012. Print.
Melnyk, George. *One Hundred Years of Canadian Cinema*. Toronto: University of Toronto Press Inc., 2004. Print.
"Mission and Highlights." The National Film Board of Canada. Web. Accessed April 12, 2015. http://onf-nfb.gc.ca/en/about-the-nfb/organization/mandate.
Moore, Paul. "The Flow of Amusement: The First Year of Cinema in the Red River Valley." In *Beyond the Border: Tensions Across the 49th Parallel*, ed. Conway and Pasch. McGill-Queen's University Press, 2013. Print.
Morris, Peter. *Embattled Shadows: A History of Canadian Cinema, 1895–1939*. Montreal and Kingston: McGill-Queen's University Press, 1978. Print.
Morris, Peter, Ted Magder, and Piers Handling. "The History of the Canadian Film Industry." *The Canadian Encyclopedia*. Historica Canada. Toronto:

January 10, 2012. Web. Accessed March 25, 2015. http://www.thecanadianencyclopedia.ca/en/article/the-history-of-film-in-canada.

Moviegoer. The (aka, John Grierson), 'Flaherty's Poetic Moana'. New York: New York Sun, February 8, 1926. Print.

Nelson, Joyce. "The Curse of Nations." In *The Colonized Eye: Rethinking the Grierson Legend*. Toronto: Between the Lines, 1988. Print.

Nichols, Bill. *Introduction to Documentary* (3rd ed.). Bloomington: Indiana University Press, 2017. Print.

Nikkel, Chris. "Arctic Canada Caught on 1919 Silent Film." *BBC News*, 21 January 2012. Web. Accessed June 16, 2015. http://www.bbc.com/news/magazine-16648847.

Renov, Michael, ed. "Introduction: The Truth about Non-Fiction." In *Theorizing Documentary*. Los Angeles: The American Film Institute, 1993. Print.

Roberts, Jerry. *The Complete History of American Film Criticism*. Santa Monica, CA: Santa Monica Press, 2010. Print.

Rose, Mandy. "Making Publics: Documentary as Do-it-with-others Citizenship." In *DIY Citizenship: Critical Making and Social Media*, ed. M. Boler and M. Ratto. Boston: MIT Press, 2014. Print.

Rosenthal, Alan. *Challenges for Documentary*. Berkeley: University of California Press, 1988. Print.

Snyder, Robert L. *Pare Lorentz and the Documentary Film*. Norman: University of Oklahoma Press, 1968. Print.

"The 1940s: The NFB." *The National Film Board of Canada*. Web. Accessed July 6, 2015. https://www.nfb.ca/history/1940-1949.

The Bioscope. London, January 20, 1910. Print.

"The History of Ukrainians in Canada." *TUGG*. Web. Accessed June 22, 2015. http://www.torugg.org/History/history_of_ukrainians_in_canada.html#The.

"The Romance of the Far Fur Country." *Wikipedia, The Free Encyclopedia*. Web. Accessed March 13, 2015. http://en.wikipedia.org/wiki/The_Romance_of_the_Far_Fur_Country.

Urban, Charles. *The Cinematograph in Science, Education, and Matters of State*. London: The Charles Urban Trading Company, April, 1907. Web. Accessed February 12, 2018. http://www.cineressources.net/consultationPdf/web/o000/332.pdf.

Véronneau, Pierre. "The Cinema of Quebec." In *The Canadian Encyclopedia*. Historica Canada. Toronto. Web. Accessed June 8, 2015. http://www.thecanadianencyclopedia.ca/en/article/the-history-of-film-in-canada.

Vertov, Dziga. *Kino-Eye: The Writings of Dziga Vertov*. Edited by Annette Michelson and Translated by Kevin O'Brien. Berkeley: University of California Press, 1984. Print.

Waddington, Damer. "Introduction." In *Panoramas, Magic Lanterns and Cinemas*. Channel Islands, NJ: Tocan Books, 2003. Print.
Winston, Brian. "Ethics." In *New Challenges for Documentary*, ed. Alan Rosenthal. Berkeley, Los Angeles, London: University of California Press, 1988.
———. *Claiming the Real: The Griersonian Documentary and Its Legitimations*. London: British Film Institute, 1995. Print.
Wise, Wyndham. *Take One's Essential Guide to Canadian Film*. Toronto: University of Toronto Press, 2001. Print.
Zhang, Haihua, and Geoffery Baker. *Think Like Chinese*. Sydney: The Federation Press, 2008. Print.

Filmography

Arrival of CPR Express at Winnipeg, Director, James S. Freer. No Production Company, 1897.
Arrivée d'un train en gare de La Ciotat, La, Directors, Auguste Lumière and Louis Lumière. Lumière and Company, 1896.
Canadian Contingent at the Jubilee, Director, James S. Freer. No Production Company, 1898.
Changing Guards at St. James Palace, Director, James S. Freer. No Production Company, 1898.
Churchill's Island, Director, Stuart Legg. National Film Board of Canada, 1941.
Drifters, Director, John Grierson. Empire Marketing Board. 1929.
Farm Boys' Camp, Regina, Director, Unknown. Saskatchewan Bureau of Publications, 1924.
Farmers of the Prairies, Director, None Assigned. Canadian Government Motion Picture Bureau. 1940.
Germany: Key to Europe, Director, Nicholas Balla, National Film Board of Canada, 1953.
Heritage, Director, J. Booth Scott. Canadian Government Motion Picture Bureau. 1939.
Hog Raising in Saskatchewan, Director, Unknown. Saskatchewan Bureau of Publications, 1924.
Moana, Director, Robert Flaherty. Famous Players-Lasky Corporation, 1926.
Nanook of the North, Director, Robert Flaherty. Les Frères Revillon, 1922.
Newfoundland Project (aka *The Fogo Film Project*), Director, Colin Low. National Film Board of Canada, 1967–1968.
Saskatchewan's War on the Grasshopper, Director, Unknown. Saskatchewan Bureau of Publications, 1924.
Sortie des Usines Lumière à Lyon, La, Directors, Auguste Lumière and Louis Lumière. Lumière and Company, 1895. All Three Versions of *La Sortie des*

Usines Lumière à Lyon Are Contained in This Video Link: https://youtu.be/DEQeIRLxaM4.
Ten Years in Manitoba, Director, James S. Freer. No Production Company, 1897.
Plow that Broke the Plains, The, Director, Pare Lorentz. Resettlement Administration, 1936.
Romance of the Far Fur Country, The, Director, H.M. Wyckoff. The Hudson's Bay Company, 1920.
War for Men's Minds, The, Director, Stuart Legg. National Film Board of Canada, 1943.
Walking Troubles of Organic Hemiplegy, Director, Gheorghe Marinescu. No Production Company, 1898.
Working Mothers, Director, Kathleen Shannon. National Film Board of Canada, 1974–1975.

CHAPTER 3

Methods and Approaches to Documentary Influence

3.1 Introduction

There exists much scholarship on documentary theory, covering a wide range of styles, approaches, modes, and technical and artistic use, but this chapter will focus only on those theories that relate directly to the genre's ability to have an impact on social change. Specifically, this chapter will examine Joris Ivens and his pioneering approach to social issue documentary filmmaking, the genre's claim to the real and its representation of the truth, how it engages audiences and stylistic approaches employing community involvement and participatory methods that yield the best results in influencing the changemaker.

The aim of this chapter is to identify the specific documentary theories that have demonstrated success in amplifying the documentary's ability to influence—and achieve—social change. Specifically, this chapter addresses the research questions posed in the introduction related to how the documentary informs and influences social change by looking at how certain approaches represent the truth, engage audiences, and yield measureable impact.

M. Terry, *The Geo-Doc*, Palgrave Studies in Media and Environmental Communication,
https://doi.org/10.1007/978-3-030-32508-4_3

3.2 Joris Ivens, 1898 to 1989

As the documentary form was finding its legs as an influential tool of persuasion in the early twentieth century, one early filmmaker was experimenting with the medium in style and content with an eye towards creating social change. Dutch documentarian Joris Ivens used film as an intervention in the socio-political process directly to correct what he saw as social injustices throughout the world. In fact, by the end of his illustrious career, he had made a documentary film on every inhabited continent (Waugh, 25), covering such issues as the socio-economic folly of reclaiming farmland from the bay of Zuiderzee in the Netherlands (*New Earth*, 1932), a miner's strike in Belgium (*Misère au Borinage*, 1934), the struggle to defeat fascism during the Spanish Civil War (*The Spanish Earth*, 1937), and the plight of Chinese refugees under attack by the Japanese (*The 400 Million*, 1939). His pioneering activist approach to documentary filmmaking earned him the title of "the most successful radical documentarist in the 1930s" (Hogenkamp, 175).

As we examined in the previous chapter, in the late twenties and early thirties, fellow documentary pioneers John Grierson and Pare Lorentz were carving out their own theoretical approaches to using documentary film as an instrument of social change—specifically, Grierson employed a didactic, educational tone, while Lorentz adopted a more poetic, propagandistic style. As José Manuel Costa argues in his essay "Joris Ivens and the Documentary Project," Ivens took an even more aggressive approach:

> (W)e cannot but acknowledge that Joris Ivens was in fact one of the main proponents of the (documentary) genre—an inventor. Throughout the whole decade of the thirties, in parallel with other genuine but different approaches (mainly, the Grierson approach and eventually the Pare Lorentz contribution), he led the way and progressively helped to define the concept (of using film for social change). Moreover, his specific path was the one which most clearly embodied the shaping process of that concept from the late twenties to the late thirties, when in many ways it finally reached the representational mode that we still call Documentary. All those three matrices (Ivens, Grierson, and Lorentz) were responsible for a common pattern of artistic search and social concern that has been definitely associated with the genre (the latter being what we could call a pattern of social productivity). But neither the Grierson public education approach nor the lyric Rooseveltian approach of Lorentz identified as clearly as did Joris Ivens with the very boundaries of that process and with the process itself—that is, the

> assumption of the avant-garde spirit and its progressive assimilation into a social, political, and historical intervention. (Costa, 17–18)

Long before the process was made popular by Colin Low, Ivens was using the participatory mode not just to tell a story, but to accelerate progressive social change for the people he was documenting. In making *Misère au Borinage* with Henri Storck, the two filmmakers collaborated with the Belgian miners in their protest for better living conditions.

> Ivens and Storck collaborated not with the government, or with the police, but with the very people whose misery no government had yet addressed, let alone eliminated. Their participatory involvement helped generate the very qualities they sought to document, not as spectacle to fascinate aesthetically and subdue politically but as activism to engage aesthetically and transform politically. (Nichols, "People," 150)

One of Ivens' most cited examples of this interventionist technique is his film *The Spanish Earth* (1937) in which the filmmaker went to the front lines himself to capture war footage. While in the middle of the chaos of war, Ivens soon learned that the structure and order with which he had previously made films was not possible. "The enemy is the co-director" (Costa, 19), said Ivens of his frustration in attempting to produce a film on the battlefield. He reiterated this sentiment in the film itself with the line "Men cannot act in front of the camera in the presence of death" (*The Spanish Earth*, 1937). Without the control that previous documentaries enjoyed in their creation, a new form of documentary was introduced.

> Behind its very pragmatic lesson on the filming of war, the practical and conceptual experience of *The Spanish Earth* can in fact be seen, metonymically, as a decisive point in the larger development of Documentary. By now, for the first time, Documentary was taking shape as a new form, where the personal authorship itself, no matter how strong it was, worked and expressed itself under new rules, demanding a new analytical frame. (Costa, 20)

An added dimension of realism or truth-telling was introduced into the documentary form, capturing images that were not scripted or even anticipated. In Thomas Waugh's seminal work on Ivens, *The Conscience of Cinema: The Works of Joris Ivens 1912–1989*, *The Spanish Earth* represents

Ivens' status as the "chief pioneer and standard-bearer" of radical documentary filmmaking.

> (*The Spanish Earth*) is the definitive model for the 'international solidarity' genre, in which militants from the First and Second Worlds used film to champion each new front of revolutionary struggle … It is also the model for the more utopian genre in which the construction of each new emerging revolutionary society is celebrated and offered for inspiration for those still struggling under capitalism. (Waugh, 195)

One of the theoretical approaches Ivens introduced in this film to augment the depiction of reality in documentary film was his use of sound. In using available sound in the field with all the challenges and shaky results that practice entails, the audio component of the film supports the visual authenticity in ways that convey a more honest and truthful representation of reality than other documentaries of the time that presented audio and narration carefully created in studios.

The theoretical approach to authenticity that Ivens' use of field sound represents paved the way for a style to be followed by documentarians in the years to come:

> *Spanish Earth*, finally, has a central place within the evolution of the documentary form, aside from its strategic ideological position. It defines prototypically the formal and technical challenges of the 30-year heyday of the classical sound documentary, 1930 to 1960, in particular its first decade. It confronts, with still exemplary resourcefulness, the problems of sound and narration; the temptation to imitate the model of Hollywood fiction with mise-en-scène, individual characterisation, and narrative line; the catch-22's of distribution, accessibility, and ideology; the possibilities of compilation and historical reconstruction, and of improvisation and spontaneity. (Waugh, 196)

As significant as Ivens' contributions were to the documentary film and its relationship to social change, some of his pioneering theories were met with criticism. In making *Komsomol* (1932) for the Russians, Ivens unapologetically staged certain scenes. In the days of *kinopravda*, this was a crime against documentary film.

> The reconstruction of events in documentary films was a disputed issue. The Soviet documentarist Dziga Vertov … was the leading representative of the

> non-intervention school. His credo was that life had to be caught unaware of the camera. For this reason, Vertov's supporters in the Soviet Union had criticized an earlier film by Ivens, *Komsomol* (1932) about the construction of a blast furnace complex in Magnitogorsk. The film ended with a 'storm night' ... except that the one shown in *Komsomol* was staged by Ivens and his crew. The Dutchman felt justified in doing this because this was the only way to get the right shots that would faithfully reflect the 'integrity and enthusiasm' of these storm nights. (Hogenkamp, 175)

The purists of *kinopravada* and later *cinéma vérité* continue to condemn scenes of reconstruction to this day—most notably, the staged scenes of reconstruction in *The Thin Blue Line* disqualified it from being eligible for the Best Documentary Feature category in the Academy Awards in 1988 (Jaffe, 51)—but many other documentarians (Flaherty, Morris, Moore, most notably) have used this technique to great advantage in their storytelling without sacrificing credibility. Still, the question of truth-telling as a theoretical absolute in documentary filmmaking continues to be debated today, long after Joris Ivens' overt use of it as a documentary style in the 1930s.

3.3 The Issue of Truth-Telling

Since the first non-fiction film was presented, definitions of the documentary have taken many forms, many of them addressing the genre's ability to convey the "truth" and capture reality as we see it. One of the early definitions of the documentary film genre in this regard is provided by John Grierson who, in 1933, described it as "a creative treatment of actuality" (Grierson, *Cinema Quarterly*, 1933). Scholars (Winston, Nichols, Renov) argue over how accurately this definition reflects a genre that lays claims to truth-telling when there are many opportunities to stage and otherwise misrepresent the truth. Robert Enright sees a conflict with the Griersonian definition explaining the contentious issue with this succinct statement: "Which part of the definition you chose to focus on would determine the kind of filmmaker you were" (Enright, 15). The word "creative" seems to be the lightning rod for those who question the truth-telling claims of the documentary in Grierson's definition. Robert Flaherty's landmark documentary feature *Nanook of the North* (1922) came under criticism when it was learned that certain scenes and events in the film were staged by Flaherty who defended himself in the now famous quote, "Sometimes you have to lie in order to tell the truth" (Pearson, Simpson, 220).

The conflict between representing reality and the artistic interpretation of that reality is at the core of the question "Is what we're seeing the *truth*?" In 1929, Russian filmmaker Dziga Vertov described his documentary films as "kinopravda" (Russian for "film truth"). Film scholar Bill Nichols explains Vertov's vision and commitment to representing the truth in his films:

> Vertov … eschewed all forms of scripting, staging, acting, or re-enacting … Vertov wanted to catch life raw-handed and then to assemble from it a vision of the new society in the process of emergence … His own term for the cinema, *kinopravda* (film truth), insisted on a radical break with all forms of theatrical, literary structure for film: these forms depended on narrative structures that crippled the potential of cinema to help construct a new visual reality and, with it, a new social reality. (Nichols, 217)

Vertov scholar Joshua Malitsky echoes this assessment citing that Vertov's goal with his early newsreel series was with the audience in mind using "film truth" as an effective way of influencing social change.

> Vertov's cinema sought to work with the viewer. He required a level of fierce sensorial, intellectual, and critical attentiveness—the kind of subject, he believed, required to properly participate in revolutionary transformation. (Malitsky, 93)

Vertov's cinematic engagement with people was not restricted to audiences. To achieve a comprehensive level of social change, he believed in making the filmmaking process accessible to everyone, according to another Vertov scholar Seth Feldman: "Vertov proposed … that Soviet films be shot by large numbers of ordinary citizens acting as film scouts, edited collectively and exchanged in a vast nationwide network" (Feldman, 26) as a means of bringing "the camera into the nooks and crannies of daily life" (Feldman, 26). This experimental technique was used in his early feature documentaries *Kinoglaz* (*Cinema Eye—Life Caught Unawares*, 1924), *Stride, Soviet*! (1925), and *A Sixth Part of the World* (1926).

Catching life "unaware" became synonymous with representing truth with Vertov. Patricia Aufderheide supports this perspective extending it to

Vertov's preference to the elimination of fiction film: "He believed that documentary was the perfect medium for revolution, that not only should it flourish but that fiction film be extinguished as a denial of the capacities of the form" (Aufderheide, 57). *Kinopravda*, to Vertov, was the only honest method of using documentary film for social change.

Films also provide a unique affordance to capturing the truth in terms of temporal considerations, according to Vertov. Vertovian scholar Annette Michelson points to Vertov's claim that the mechanical eye of the camera can see things the human eye might miss, thereby providing a greater access to visible evidence and, ultimately, the truth. What the human eye might miss at one moment, the mechanical eye will capture at the same moment, allowing sight of it by the human eye at a later time. In translating an analysis of this made by Gerard Granel, she writes:

> His work is paradoxically concrete, the original and paradigmatic instance of 'an attempt to film, in slow motion, that which has been owing to the manner in which it is perceived in natural speed, not absolutely unseen but missed by sight, subject to oversight. An attempt to approach slowly and calmly that original intensity which is not given in appearance, but from which things and processes have nonetheless in turn derived. (Michelson, xix)

In the following chapter, we will examine how these considerations of time distortion—both fast and slow, now and later—can be used to develop implicit narratives, a key component of truth representation that provides a fuller understanding of the social issue profiled in a documentary film for the changemaker.

Inspired by Vertov's dedication to representing the truth, French filmmakers Jean Rouch and Edgar Morin coined the term "cinéma vérité" (French for *kinopravda*) in the 1960s to describe a specific brand of documentary film. While Vertov used his "film truth" term to describe all cinema, Rouch and Morin used theirs to describe a specific style of documentary:

> Jean Rouch and Edgar Morin … named their new form of documentary filmmaking cinéma vérité … as a *type* (or mode) of documentary, rather than as an all-inclusive category. A term that had begun with Vertov as the definition of all true cinema became associated not only with the more delimited area of one genre, documentary, but also with the further delimited mode of participatory documentary. (Nichols, 218)

Another definition of the documentary addressing this aspect of "truth-telling" was provided by Michael Renov in 1993. It is similar to Grierson's in that he defines documentary film as the "artful reshaping of the historical world" (Renov, 11), and, as such, the documentary "has struggled to find its place within the supposed conflict between truth and beauty" (Renov, 11). The "beauty" referred to in Renov's assessment relates more to the visual presentation of the content in a documentary film rather than to its literal definition. Sometimes the truth of a documentary can be quite ugly, and packaging it to look beautiful can indeed be a challenge, but Renov is aware of this when he uses the word "supposed." There are several methods, styles and production tools such as narration, animation, even dramatic re-creations, that can be artful and beautiful while still contributing to and underscoring the truth, as I explained earlier in analysing Flaherty's statement about the need to lie to tell the truth.

In his essay, "Toward a Poetics of Documentary," Renov points out that documentary style, not just content, can be a testament to the truth being presented. The shaky-cam and underlit production flaws of a documentary like *Kon Tiki* (Thor Heyerdahl's 1950 documentation of his 4500-mile sea voyage on a raft) are "equally witness to its authenticity" (Renov, 23). Not all scholars, however, are convinced that the documentary can present the truth in absolute terms, despite the assistance of technical and artistic support. Noted documentary scholar Brian Winston referred to Grierson's definition to suggest that any objective truth claimed by the documentary is impossible:

> Surely, no 'actuality' (that is, evidence and witness) can remain after all this brilliant interventionist 'creative treatment' (that is, artistic and dramatic structuring) has gone on. Grierson's enterprise was too self-contradictory to sustain any claims on the real, and renders the term 'documentary' meaningless. (Winston, *Claiming the Real*, 59)

Bill Nichols offers a perspective that separates the objectivity of a journalist from the content in documentary films, asserting that the "creative treatment of (filmed) actuality" cannot be avoided:

> Documentaries always were forms of re-presentation, never clear windows onto 'reality'; the film-maker was always a participant-witness and an active fabricator of meaning, a producer of cinematic discourse rather than a neutral or all-knowing reporter of the way things truly are. (Nichols, 19)

He is right, of course, that there can never be a "clear window" in the process of production and the representation of reality in documentary film—with the possible exception of real-time surveillance—even then there is a creative decision being made as to where the camera is positioned. The technology and the temporal access to the reality at the time of shooting prevent this; however, certain "creative treatments" such as animations and narration can help provide clarity.

In his essay, "We Aren't Sorry for this Interruption …," Steve James understands the documentary's function as follows: "Documentary has two main objectives: to reveal the world and/or to change the world" (James, 58). In addressing the issue of truth-telling, in particular, James explains the importance of activist follow-up by the filmmaker:

> (The) duality of purpose between film and outreach works very well. It allows the films to pursue a creative and journalistic "truth", while being used in a very direct way to effect progressive social change … the ultimate effectiveness of the latter is due in large part to the integrity of the former. (James, 58)

The journalistic approach to conveying the truth through the direct and indirect personas of the filmmaker allows the documentary to become an effective communications tool in the process of social change. By journalistic, I mean the objective reporting of information both observed and stated by interview subjects. Starting in Canada in 1897 with its first filmmaker a journalist by trade, we see the documentary's first general use in this regard, and later, it develops into Grierson's more focused use based on his ideological belief that the National Film Board of Canada ought to serve as the state's cultural tool, specifically, "to bring about positive social change" (Rosenthal, 6).

As examined in Chap. 2, one of the first uses of non-fiction film was for education, often in the field of science. These "animated lectures," pioneered by the Lumière brothers and James Freer, frequently included the filmmaker who provided live narration throughout the screening in addition to introductory remarks and answering questions following the screening. It was very much a teacher's tool and was used this way to great extent and success by physicians Gheorghes Marinescu and Jean Louis-Doyen at the end of the nineteenth century. Any truths obfuscated or omitted by the journalist-filmmakers in their documentary films could be added in their live presentations.

While the testimony of interview subjects may be suspect and the decision of which content to include and to discard by the filmmaker may be questioned, the representation of scientific content in documentary film introduces a more trustworthy subject rooted in objective facts, and not subjective memories and opinions. This theory is articulated quite well by Kirsten Ostherr in her essay "Animating Informatics: Scientific Discovery through Documentary Film." There exists a "dispassionate" element to the representation of scientific data in documentary film allowing for an increased level of perceived truth-telling and not requiring the same degree of emotional engagement that social issue documentaries so often require.

> By viewing scientific images as powerful rhetorical expressions that participate in the techniques and goals of the documentary tradition, we may begin to mobilize their persuasive power to enable a fuller understanding. (Ostherr, 28)

Whenever a representation is made in photography or film, several choices are being made by the photographer or filmmaker: lighting, framing, setting, interview subject and their respective testimony, and many more. This "creative" treatment of the truth compromises the essence of the truth to some degree making absolute truth impossible to represent in documentary film. This concept is best described by Linda Williams in her chapter "Mirrors Without Memories: Truth, History and the New Documentary": "We do better to define documentary not as an essence of truth but as a set of strategies designed to choose from among a horizon of relative and contingent truths" (Williams, 799). The intent of these strategies is to present to audiences various truths related to the documentary's subject in order first to inform, and second to influence through engagement.

3.4 Audience Engagement

Once a relationship of trust is established between a documentary film and its audience through the presentation of what it perceives to be truthful content, the next step is to get that audience to act. A key component in achieving that goal is to engage them emotionally. This is a widely held theory with many investigations by scholars and practitioners as to how the documentary does this most effectively.

Film theorist Carl Plantinga explains how documentary film achieves this goal of emotional engagement:

> In movie spectatorship, as in the rest of life, the repetition of elicited emotions and judgements may solidify ways of thinking and feeling. It is through the elicitation of emotion in relation to moral and ideological judgement that a film may have its most significant ideological force. (Plantinga, 203)

Documentaries achieve this by depicting more than one example of a social issue. The repetition of these examples, together with the emotional impact associated with them, builds and reinforces the audience's emotional engagement with the message of the documentary. This emotional engagement is not exclusively aimed at a viewing audience. According to Bill Nichols, there exists a three-way relationship in documentary film—between the filmmaker, the subjects of the documentary, and the audience (Nichols, *Introduction*, 59). He identifies three common formulations of this theory, the first of which is described as "I speak about them to you." The filmmaker, Nichols argues, takes on a "personal persona," either directly or indirectly through a "surrogate," a technique first popularized by many of John Grierson's early films with the NFB. The surrogate is often the "voice-of-God" narrator, as seen in such nature documentaries as *March of the Penguins* (2005) and *Yosemite: The Fate of Heaven* (1988). The narrator is often unseen to the audience, but not always. If the narrator is recognized as being an expert or an authority in the field of the documentary subject or is perhaps known for holding some significant credentials in journalism, the filmmaker might present them on camera in a Brechtian address directly to the audience. Either technique provides an authoritarian voice for the audience to underscore the credibility of the content and extend the film's emotional engagement with its audience.

A further extension of this involves the filmmakers themselves appearing on camera as a guide through the story. In this participatory approach (the specific mode of which will be examined later in this chapter), the audience joins the filmmaker on their journey through the documentary's investigation. Nichols indicates that this technique "shifts the emphasis from persuasion to expression" (Nichols, *Introduction*, 60). In addition to the content itself, the audience now needs to evaluate the credibility of the filmmaker and the way they express themselves and their investigation. This becomes a risky proposition when documentarians like Michael Moore are called out for cherry-picking, at best, and falsifying, at worst, documentary content in the same "the-end-justifies-the-means" way Flaherty was found to have done by staging certain events in *Nanook of the North*.

There are also times when the audience is left with no narrator, on-screen host or filmmaker guide. With these films, the audience is left on their own to evaluate the story and to be engaged without narrative assistance; however, the documentary increases its chances of emotional engagement with the additional support of the filmmaker's voice or their surrogate, provided they are perceived to be reliable and ethical, in short, trustworthy.

Achieving an audience's emotional engagement is only part of the equation that leads to successfully influencing those who can affect social change. In Patricia Aufderheide's paper delivered at the 2015 Visible Evidence conference in Toronto, she describes a strategy for engaging an audience emotionally. The strategy represents the "dimensions of impact" (see Fig. 3.1) any documentary film project needs to have to achieve social change, should that be its goal. The process was created in 2009 by the Fledgling Fund, a US-based foundation that funds social justice documentaries, and has been used by scholars studying documentary impact ever since. It begins with a *compelling story*. That will lead to *awareness* of the film's issue and if presented successfully will lead to *audience engagement*. This, in turn, leads to a *stronger movement* and, ultimately, *social change*.

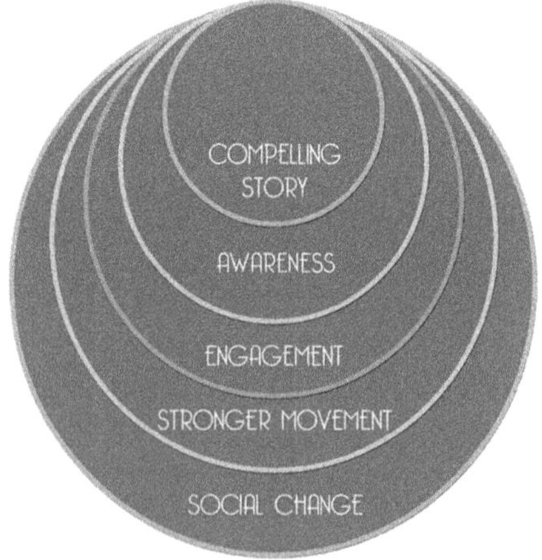

Fig. 3.1 "Dimensions of Impact". New York: *The Fledgling Fund*, 2008. Barrett, Diana and Sheila Leddy. http://www.thefledglingfund.org/resources/impact

In her essay "Political Mimesis," Jane M. Gaines explains how when a documentary audience is sufficiently engaged emotionally, members are more likely to act on the issue being presented:

> The reason for the documentary to advance political goals is that its aesthetic of similarity establishes a continuity between the world of the screen and the world of the audience, where the ideal viewer is poised to intervene in a world that so closely resembles the one represented on the screen. (Gaines, 92)

For social change to be realized through a documentary film, audiences relating to the message of a showcased issue they either were familiar with or one that was completely foreign to them need to be emotionally engaged. This applies both to the general public who can be moved to take action outside the cinema and to the policymakers who use the content of the documentary to assist them in advancing resolutions, recommendations, programmes, laws, and policies that propose and legislate progressive social change.

Today's impact-funding community echoes this philosophy. BRITDOC, a non-profit organization dedicated to "enabling" social issue documentaries through education and funding programmes, advises its clients of the importance of emotional engagement with their audiences:

> Unlike shorter forms such as news and social media, long form documentary takes the time to build empathy more deeply, involving audiences directly and immersing them fully in the situation of others, prompting them to engage and act. ("The Power of Film," *Impact Guide*)

While the debate continues over the documentarian's right to "creatively" or "artfully" represent reality and their obligation to present it without distortion or deception, there are other definitions of the documentary and its construct that more closely represent the subject of this investigation. One of them relates to a filmmaker's understanding of their audience. If the subject of a documentary is too technical, obscure or otherwise unfamiliar to an audience, they may disconnect from the film and its messages. An artistic strategy to prevent this from happening is to translate the content of the film into easy-to-understand conceptual clichés.

In Kara Keeling's book *The Witch's Flight: The Cinematic, the Black Femme and the Image of Common Sense*, she explains film's "ability to tear a 'real image from clichés'" (Keeling, 14). Keeling describes a sensory-motor apparatus at work when films are watched that allows for continuous movement of audience engagement when the cliché is employed. When this movement is sidetracked by unfamiliar (non-cliché) images and messages, a different kind of movement takes place—thinking. It is therefore essential for the social issue documentarian to consider the audience in telling a story that is both familiar enough to keep their attention and "non-cliché" at times to stimulate thought and, eventually, action.

3.5 The Four Functions of Documentary Practice

In his essay, "Towards a Poetics of Documentary," Michael Renov identifies four[1] fundamental functions attributable to documentary practice: to record, reveal, or preserve; to persuade or promote; to analyse or interrogate; to express (Renov, 25). These categories are essential to understanding how and why the documentary makes claims to the real, how it influences changemakers, how it critically challenges social injustices, and finally, how it affects audiences it targets and with which it engages.

In Patricia Aufderheide's succinct summary of the genre, *Documentary Film: A Very Short Introduction*, she cites the four "founding fathers" of the documentary as first identified in 1971 by Erik Barnouw in his book, simply titled, *Documentary*: "(Robert) Flaherty, the Explorer; Dziga Vertov, the Reporter; young Joris Ivens, the Painter; and (John) Grierson, the Advocate" (Aufderheide, 130). The work of these early pioneers can be seen to represent their preferred Renovian function of documentary practice: with his heralded ethnographic account of the life of an Inuit and his family, *Nanook of the North*, Flaherty the Explorer embodies the function "to record, preserve or to reveal" previously unknown cultures in previously unknown parts of the world; with his depictions of Russian class structures (*Man with a Movie Camera*) and colonialism (*A Sixth Part of the World*); Vertov the Reporter fulfils Renov's function "to analyze or interrogate" as he questions poverty in Moscow and the motives of the Russian State Board of Trade for introducing audio speeches of Lenin on vinyl records to Sami tribes, the Soviet Arctic native people; Ivens the

[1] Michael Renov's four functions of documentary practice are considered the primary purposes of documentary film, but in Chap. 4, we will examine a fifth function provided by Chilean anthropologist, documentary filmmaker, and media scholar, Juan Francisco Salazar: the Anticipatory Mode, specific to ecocinema documentary theory and practice.

Painter uses a photographic palette in *Rain* to structure visually attractive scenes to express meaning and emotion; and, finally, Grierson the Advocate represents the function to promote and persuade with his government mandate to use documentary film "to bring about positive social change" (Rosenthal, 6) at the state-run National Film Board of Canada.

Renov's four functions of documentary film can be mapped back to its origins suggesting an inherent quality of the documentary film as a persuasive communications tool, a cumulative function of all four.

3.6 Production Methodologies to Promote Social Change

As documentary film evolved in the middle of the twentieth century, several experimental methods of production, distribution, and exhibition were used to accelerate and ameliorate the genre's ability to affect social change. Bill Nichols introduces six "modes" of documentary filmmaking that define the styles of the contemporary social issue documentary.

One of these modes, participatory, has demonstrated a measurable effectiveness in influencing changemakers. In addition to Nichols and Winston, proponents of this theory include Janine Marchessault (*Mirror Machine: Video and Identity*) and Brian Rusted (*The Politics of Distribution and Counterpublics*).

Along with Ivens, another early documented use of this approach as a methodology was by a Quebec priest, Father Albert Tessier, to involve rural communities in the filmmaking process (Véronneau, *The Cinema of Quebec*). This mode also extended into other forms of involvement. In addition to contributing to the film's production and audience participation, this inclusive documentary technique extended to the film's distribution. A grass-roots release of these documentaries became part of the process where screenings were held within the profiled communities, often to those individuals showcased in the films. Film scholar Brian Rusted explains this process in his essay, *The Politics of Distribution and Counterpublics*:

> (By) imbuing documentary cinema with radical participatory impulses at the levels of both production and distribution, (this approach created) a revolution in community organizing and development of communications which continues to this day. (Rusted, 405)

The documentary film had adopted a new voice in its storytelling method—the interview subject—amplifying and refining messages necessary to stimulate progressive social change.

Today, the use of participatory filmmaking in documentary is adapting to new technology and affordances in the digital domain. This will be explored to greater extent in Chap. 5, but its expanded use provides documentary audiences more immediate and direct opportunities for activism, as Siobhan O'Flynn explains in her paper "Documentary's Metamorphic Form: Webdoc, Interactive, Transmedia, Participatory and Beyond":

> (T)he impact of social media and the rise of participatory strategies of engagement have positioned audiences as collaborators and creators who can expect an immediacy of response and the opportunity for agency. (O'Flynn, 152)

The "participants" are engaging more as "collaborators" by directly contributing user-generated content to documentary projects, amplifying their voice, and reaching many more viewers through social media channels. Among this expanded audience is the changemaker. They can be reached more easily through a digital connection, as we will explore in Chap. 5, and sometimes even serve as collaborators themselves.

3.7 The Seven Modes of Documentary Film

As alluded to previously in this chapter, Bill Nichols describes a style of documentary filmmaking that has seen much success in audience engagement and social change outcomes: the participatory mode, one of the seven modes of documentary filmmaking he identifies (the other six being *expository*, *poetic*, *observational*, *reflexive*, *performative*, and *interactive*[2]). He describes this inclusive mode as embracing not just the subject(s) of the documentary, but its audience as well: "This mode inflects the *I speak about them to you* formulation into something that is closer to *I speak with*

[2] The interactive mode was added by Nichols in his *Introduction to Documentary* (Third Edition) published in 2017.

them for us ... as the filmmaker's interactions give us a distinctive window onto a particular portion of our world" (Nichols, *Introduction*, 179–180). While proven to be a successful technique with the Fogo Process, Brian Winston argues that, here, the filmmaker has introduced a moral responsibility required of the filmmaker as to how much to show through that window, a responsibility seldom assumed. In his essay, "The Tradition of the Victim in Griersonian Documentary," Winston asserts that when working in this mode, the filmmaker fails to consider the "vulnerable individual." In fact, the profiled family members in Tanya Ballantyne's *The Things I Cannot Change* were harshly ridiculed by their neighbours following the film's public airing on the Canadian Broadcasting Corporation's national television network causing them to suffer "further alienation and ostracism, forcing them to move" (Waugh et al., 11).

Winston further illustrates the filmmaker's culpability in this process: "By choosing victims ... the moral and ethical implications ... are not only ignored, they are dismissed as infringements of filmmakers' freedoms" (Winston, *New Challenges*, 276).

The significance of this mode extends beyond safeguarding interview subjects from public ridicule; it also offers an enhanced mode of communication to those charged with improving the lives of those profiled, as Rusted suggests. It empowers the filmmaker and the community participants to explain their situation more effectively to government officials and thus serves as a more powerful influence for social change. The partnership between filmmaker and community creates a strong and unique voice in communicating social needs to power.

The participatory mode owes a lot to the advent of a new technology introduced in the mid-1960s, the Sony Portapak. As its name suggests, the Portapak was a portable video recorder that allowed for instant viewing of recorded material, unlike its predecessor, film. Its relative ease-of-use and smaller camera unit, together with its instant playback, made the new equipment accessible to everyone, especially interview subjects. One of the best examples of using this new technology, and one of the most cited examples of the successful uses of the participatory mode to affect social change, was a collection of twenty-six films made of the people of Fogo Island in 1967 and 1968 known as the *Newfoundland Project* series (Waugh et al., 224). Inspired by the community-based approach of Father Tessier in Quebec, documentary filmmaker Colin Low employed this technique for the National Film Board of Canada series, *Challenge for Change* (CFC).

The 5000 people living in the Newfoundland fishing village on Fogo Island were slated to be relocated by the federal government in response to the dire poverty growing there at the time. Resisting this unwanted move, the villagers eagerly participated in Low's experiment. Fogo residents actually shot much of the footage themselves and were brought into the editing process where their contributions were encouraged and acted upon. As a direct result of screening these documentary films to government officials, "(the Canadian government) scrapped the relocation plan, and in its stead provided assistance and encouragement to the islanders to start a fishing cooperative and marketing board" (Jones, 163).

Film demonstrated its utility as a vital communications tool for the citizens of Canada and its government, and the participatory approach provided an enhanced form of credibility to the claims made by Fogo residents previously made only through written text. It was able to deliver a message of urgency via actual footage of Fogo islanders speaking directly to the camera about being affected adversely by the proposed government relocation—something that written texts and formal speeches had failed to deliver. In its description of the Fogo film series, the NFB describes the films as showing how the process "can be a catalyst for social change by serving as a direct means of communication" ("Introduction to Fogo Island," National Film Board of Canada website).

As previously examined, it is rare for a documentary film to demonstrate a direct impact on action that causes progressive social reform; however, the Fogo experiment is one of the rare examples of the measureable impact a documentary can have in assisting in the process. Instead of serving as an executive producer, as the government has been seen to do in its earlier partnerships with the filmmaker and the corporate sector, federal bureaucrats in this instance take on the role of the audience. Social change was instituted by the government as a result of Canadian citizens, now working with the filmmaker, communicating their social needs directly through film. They were able to change the government's plans for an unwanted move from their homes. With the earlier goals of immigration, promotion and education, we have seen the government use the filmmaker to create the potential for social change for its citizens, even though its films might be classified as examples of government propaganda; now, we see the people of Canada, at least those on Fogo Island, working in conjunction with the filmmaker to create self-authored forms of social change communication by influencing the government through personal filmed testimony.

Documentary film scholar Janine Marchessault defines this mode in terms of its function to introduce a new dimension of truth-telling to the documentary's content: "The participatory approach to documentary filmmaking is not a stylistic but an ethical engagement with the process of representation" (Marchessault, 16). The documentary film had adopted a new voice in its storytelling method—the documentary subject—adding a key partner to the triumvirate of filmmaker-government-corporate sector that emerged in Canada's early documentary years, and amplifying and refining messages necessary to stimulate progressive social change. The Fogo example re-positions the filmmaker as partner and collaborator in both the filmmaking process with interview subjects, but also as an advocate for them with the government after the filmmaking process. This targeted audience of changemakers used the films to institute a policy that best served those profiled in the Fogo film series.

Following up on the success of the Fogo Process and making use of the new Portapak video technology, was another experiment, this time with an impoverished urban community in Montreal: *VTR St-Jacques* (1969). The people of St-Jacques, a small neighbourhood in Montreal, had wanted to improve their community and formed a Citizens Committee comprising "workers, unemployed, students, housewives and welfare recipients" (Klein, *VTR St-Jacques*). One of the first projects of the committee was to establish a medical clinic for its neighbourhood, "after petitioning in vain to the authorities for better health care" (Klein, *VTR St-Jacques*). There was a need to communicate the existence of this clinic to the community, but, as the film suggests, access to the mainstream media is not easily available, "especially to the poor" (Klein, *VTR St-Jacques*). As a result, the CFC filmmakers asked this question: "What could happen if the people had the technology of communications in their own hands?" (Klein, *VTR St-Jacques*), and set out to document the experiment of putting the new portable video technology in the hands of the needy for the purpose of communicating better to the authorities.

After a brief training session with CFC filmmakers, the members of the St-Jacques Citizens Committee began filming each other. One of the interesting results of this training is how quickly the inexperienced filmmakers adapted to all aspects of filmmaking, not just camera operation. In one scene, the committee watches one of its members being interviewed about unemployment in the neighbourhood. He states that one of the reasons for this problem is that women are entering the workforce. They should stay home, he says. The interviewer asks if they are taking work away from men, and he emphatically agrees. The committee watching this interview instantly become broadcast editors. They say they should not include this interview in what they plan to present publicly because "he represents the old way of thinking." Another member reminds the committee that it is "his" opinion, and all voices of the neighbourhood need to be heard. Editing, the film says, is the hardest part of the process.

It is also one of the most surprising aspects of participation in the documentary filmmaking process for the inexperienced Citizens Committee members. There was some concern among journalists that the opportunity afforded to the St-Jacques community might result in the tail wagging the dog, transforming their documentary from an objective report into a subjective editorial. While this was clearly the committee's intent, the spectre of propaganda masquerading as responsible journalism loomed. Canadian broadcast journalist Patrick Watson expressed this concern in a 1970 article in the national arts magazine, *artscanada*, "It's one thing to have people participate in planning and even shooting a film about their lives, quite another to involve them in the editing process" (Watson, 14).

The most intriguing objection to the entire process came from the Montreal media themselves. In the film, a press conference is held, and local journalists, dressed in formal suits, show up only to learn the committee wants to interview them on camera for the film project. This did not go over well. At one point a journalist says, "We're being filmed … I don't know why." A nervous committee member tries to explain something she did not think needed explanation. After all, journalists show up at events all the time without advance notice or permission and film those involved in the name of news. Why could the Citizens Committee not do the same? Another irate reporter, feeling ambushed, says "I come here and suddenly I'm a star. If you asked in advance, perhaps I'd agree, but this, I don't like" (Klein, *VTR St-Jacques*).

It is with incredible irony that the media stake their claim with these objections. The privileged hegemony of Canadian journalism is clearly resisting the rise of the other, the (non?) working class in this democratizing tactic afforded by the new video technology and the participatory mode of documentary filmmaking.

The portable technology certainly contributed to the community's ability to communicate better with government officials, but the two directors of the film (one on-camera, Dorothy Todd Hénaut, from the community; the other, Bonnie Sherr Klein, behind the camera, from the NFB) make an interesting point about how this new equipment impacts the documentary's ability to cause social change:

> Video equipment does not create dynamism where none is latent; it does not create action or ideas; these depend on the people who use it. Used responsively and creatively, it can accelerate perception and understanding, and therefore accelerate action. (Sherr Klein, Todd Hénaut, 5)

As we will examine in subsequent chapters, advances in technology frequently serve the goal of facilitating the documentary film's ability to communicate and influence, strengthening its power to serve as an agent of social change.

3.8 Studio D

The participatory mode, as it was developed and practised in documentary film production in Canada, often provided a voice to those who could not speak for themselves. Father Tessier provided representation to remote aboriginal communities, and Colin Low represented the remote fishing community of Fogo Island—both communities in dire need of economic improvement. Another disenfranchised community in Canada—women—took advantage of this new mode of community filmmaking within the NFB to improve their own situation.

On December 7, 1970, the Canadian government released the *Royal Commission Report on the Status of Women*, calling on all government agencies to re-assess their employment structures. In particular, Recommendations 4 and 5 specifically called on the agencies and departments of "federal, provincial and territorial governments" to be proactive in the hiring of women and for equal pay (*Royal Commission Report on the Status of Women*, 395–396). This document was a direct response to a shift towards acknowledging women as citizens and attending to their voice. As a high-profile federal agency, the NFB realized they needed to address gender equality both within its own ranks and on screen—a move the hierarchy of the NFB strongly resisted.

While working within a government agency—the NFB—and making films to present to another government entity—the legislators in Ottawa—challenges surfaced that would make the community filmmaking process problematic. The structural change at the NFB to allow more women filmmakers and more films with a distinctive voice of women was a difficult one for the NFB to make. The NFB's sexism at this time had deep roots reaching all the way back to Grierson who saw little place for women in film production, as described by Gail Vanstone in her book, *D is for Daring*.

> From its earliest days, almost without exception, the NFB had excluded women from formal filmmaking, following NFB founder John Grierson's philosophy that women should not assume prominent roles. (Vanstone, 38)

However, the political climate in Canada at the time played a key role in establishing the NFB's all-women division, Studio D. In developing programmes for the United Nation's International Women's Year in 1975, the Canadian government asked the NFB to "recognize (its) priorities" and create a women's programme that would be "really special" (Vanstone, 38). As a result, Studio D was born. Studio D was established in a "modest" location (the former janitor's room in the basement) with a small staff. Studio members found themselves in a state of "perpetual dependency" (Vanstone, 54) on the National Film Board in these early days, if not always. Studio D had a mandate of bringing about social change through its own brand of "radical liberalism" that articulated that "change is possible and that state action is an acceptable way of achieving change" for women (Vanstone, 83). In this case, the "state action" would be the films made by the women of Studio D.

Studio D's founder and head, Kathleen Shannon, worked with the *Challenge for Change* series and adopted a participatory mode approach to her documentary storytelling style. As we explored previously with the Fogo experiment, Shannon's method of seeing stories through a lens of "life experience" allowed her to establish a unique trust between female filmmaker and female interview subjects in her film series *Working Mothers*, made under the *Challenge for Change* banner. While the NFB may not have provided all the support she would have liked, its official description of the series heralds it as an important body of work in promoting social change: "(the series) offer(s) audiences a point from which to assess the gains made by women over the last two decades, and emphasiz(es) the ongoing need for social and political change" (Vanstone, 108).

The particular approach of showcasing individual and collective perspectives of women's lives with an analysis of how women were defined at the time and oppressed and repressed by socio-political structures "served as prototypes for other feminist documentarians" (Armitage et al., 44), but getting funding and green lights from the entrenched male hegemony of the NFB was a struggle. As a result, Shannon employed new techniques in meetings with them to achieve the goals of Studio D. Terre Nash, director of the Oscar-winning film *If You Love This Planet*, remembers the unique approach Shannon took to get her documentary about the dangers of nuclear warfare funded:

> We had weeks and weeks of meetings with these men and she totally unnerved them because she knitted all the way through. It was extraordinary to watch these big, powerful bureaucrats having temper tantrums, only to be met with a look of disdain from the stony face of Kathleen Shannon and the rhythmic clacking sound of knit one, purl one. … When it was clear that we had won, Kathleen looked at me and smiled and said, "Well you've got your film and I've got a new sweater!" (Nash, 38)

The controversial film was labelled propaganda by the US government, yet still went on to win the Academy Award that year in the category of Best Documentary Short. Demonstrating the strength of the film as an influential political tool, it was embraced by Canada's own Liberal Party when Prime Minister Pierre Trudeau invited the star of the film, Dr Helen Caldicott, to a policy session on international affairs. This may have "played a small but instrumental role in the genesis of Trudeau's own peace initiative" (Vanstone, 164). Whether it did or not is not as significant as the fact that a symbiotic relationship between the filmmaker and government collaborator was emerging.

At the time, a women's liberation slogan, "the personal is political" (Hanisch, *The Personal is Political*), reflected the focus of many of Studio D's films. These films represented a strengthening of the objective first realized by Colin Low and the *Challenge for Change* series in which the filmmaker was now using the medium as an enhanced communications tool and one of social activism to influence the government to enact progressive social reforms.

These new documentary styles and voices added to the growing use of the documentary as an instrument of social change, but now the filmmaker was delivering the message directly to those charged with affecting social change. Many of the films of Studio D, following the precedent set by the Fogo Process, illustrate that it is not the participatory mode alone that yields social change results most expeditiously, but rather targeting the changemaker directly as the film's intended audience and customizing style and content to best accommodate their needs.

3.9 The Direct Approach to Influencing Social Change

The lack of evidence directly linking a documentary film's subject matter and its impact on social change is not widely considered. Many believe *An Inconvenient Truth* was the reason for the surge in global climate action, when in fact the United Nations was holding climate change conferences ever since the Kyoto Protocol in 1997, nine years before the film was made. There is a perception that social issue documentaries and activist documentaries result in social change outcomes, but as Brian Winston says, "(t)he underlying assumption of most social documentaries—that they shall act as agents of reform and change—is almost never demonstrated" (Winston, *Claiming the Real*, 236). In the rare instances when there is a direct connection, there is often a common factor: presenting the film directly to the changemaker. Instead of influencing a general, public audience who may or may not take action to extend the film's influence to changemakers, documentaries like *The Newfoundland Project* that are made for and presented to those charged with affecting social change demonstrate the strongest and most consistent connection between the documentary film and social change. While naysayers insist there is seldom evidence proving a documentary film's ability to cause social change directly, this new relationship between the filmmaker and the changemaker holds promise that a direct connection between a documentary and its intended social change is not only possible, but measureable.

In a 2009 study on the influence of the documentary film among government "decision makers and policymakers" conducted by Dr George Ferreira, Programme Lead at the Ontario Ministry of Agriculture, Food and Rural Affairs, involving a series of interviews with his fellow bureaucrats, some definitive attitudes on film as a preferred and effective information source for policymakers are revealed. Here is a typical comment from a government bureaucrat who served as a research subject in the study:

> I read significant documents all the time, but it goes out of my mind as I read the next significant thing. I keep shifting several times a day … and I think the video has much more lasting power. The video in my mind has more lasting impact than a report. Maybe it's the concentration factor as

> well. … It'll take me three hours to make my way through a 50-page report … but a video can take you through the concept in 15 minutes and add a human dimension that the report cannot. (Ferreira et al., 33)

But how does the community reach the policymaker when distance is a barrier? It is often too costly and time-consuming to visit remote communities to see first-hand the social problems requiring government support, but together with the filmmaker, documentaries made in the remote communities and presented to the policymaker promise to bridge that geo-communicative gap.

In particular, the participatory film method that Joris Ivens and Father Tessier pioneered, and that Colin Low developed, is seen to be particularly valuable to policymakers with respect to understanding the social situation in remote locations. The report refers to the Fogo Process specifically, calling it an "historical antecedent" and used as a model in this study (Ferreira et al., 1, 22). Dr Ferreira's study was called *Influencing Government Decision Makers Through Facilitative Communication via Community-Produced Videos: The Case of Remote Aboriginal Communities in Northwestern Ontario, Canada*, and as we have seen from the Fogo experiment, community involvement in the documentary filmmaking process provides a context specific to the subject community, not merely the perspective of a filmmaker from the "outside."

The study focused on a series of community-based films on the economically impoverished aboriginal Keewaytinook-Okimakanak communities in Northwestern Ontario that were produced and presented over a two-year period (2003 to 2005) to twenty-two decision makers at Industry Canada and the Ontario Ministry of Agriculture, Food and Rural Affairs. The study presents five conclusive findings on how government officials view the films and how they influence their decisions:

- community-produced videos provide a highly valuable context for policymakers about communities;
- videos can be used to inform and galvanize federal staff working in the service of these communities who might not otherwise have the opportunity to visit these communities or meet their inhabitants;
- community-produced videos are a legitimate and effective way of providing qualitative data for policymaking processes;

- videos can serve as an organizing structure or event around which senior bureaucrats and politicians can form policy directives and influence other policymakers; and
- videos have the potential to influence policymakers, thereby shifting the direction of policy in response to community needs and aspirations (Ferreira et al., 1).

The report clearly attributes community-produced videos as an influential, legitimate, and effective source of data for policymakers. The particular style of the Fogo Process—the participatory mode of documentary filmmaking and presenting the films directly to changemakers—provides further evidence of the Canadian documentary's ability to affect social change among its country's government policymakers. Again, we see the filmmaker and the new community partner working together to change policy from the bottom-up. Originally, it was shown how the government used the filmmaker to create social change from the top-down.

The programme that Dr Ferreira began in 2003 established an open digital dialogue between the aboriginal Keewaytinook-Okimakanak communities in Northwestern Ontario and federal policymakers. According to Dr Ferreira, the programme never ended. After the local people were trained in filmmaking and editing, they continued to produce films on social and economic issues related to their communities for federal policymakers, adding that the native communities of Northern Ontario have produced "hundreds, if not, thousands of videos" since 2003.

Dr Ferreira's research quotes a senior programme officer for Industry Canada who summarizes his department's appreciation of documentary films in this regard:

> There's a bureaucratic barrier that doesn't allow us to travel to these remote areas. If you don't have, in your mind, a hands-on experience, face to face with the people doing the work in remote or rural areas, then you don't have a concept of what it's like. You've got to have something to provide us in Ottawa with a feel for what's going on. If you can't make the visit, you've got to think of other ways of doing it. And so we've got to think of other ways of communicating what's happening on the ground. Videos, especially the way you've produced them in conjunction with the communities, really can go a long way in filling that gap. (Ferreira et al., 167–168)

This assessment reveals the trust placed on the content and messages delivered in these films by government officials through the participatory mode process. The quoted bureaucrat articulates that the community-based method provides a credibility to policymakers that the community representing the policy to be changed is represented directly in the film, not as a mere interview subject, but as an active participant, a co-producer of the film to a degree, so that an understanding beyond what is traditionally available through text alone can be achieved.

To show first-hand the conditions in which a community is living, presented by members of that community in their own words, provides more credibility of the issue being profiled. As one federal programme officer said in Dr Ferreira's research:

> When a bureaucrat visits a (remote) community they come with their own pre-conceived notions and ideas and often, because they don't stay very long or they only see the surface, the broken down trucks all over the place, they don't get the whole story. What these videos allow the communities to do is construct the image that they see of themselves and, in a way it's actually much more honest. (Ferreira et al., 170)

This model illustrates a desire for this kind of documentary film as a reliable and "more honest" data delivery system to assess a situation that requires improvement. It also reveals a new relationship between the filmmaker and the government, whereby a particular style of documentary filmmaking is being invited by the government, now acting as eager accomplices, working with activist filmmakers to collaborate for progressive social change for its nation's citizens.

3.10 Conclusion

The participatory mode introduced a new way of perceiving documentary content, messages, and calls to action. Making documentaries in this style and presenting them directly to the changemaker seem to yield the best and most measurable results in impacting social change. Theoretical reconstructions of the genre such as this are frequently aimed at influencing an audience to action to drive members of society to be the catalysts of their own progressive social change. Renov's four functions of documentary filmmaking—to persuade, to interrogate, to express, and to preserve—all contribute to influence social change using the participatory mode. One sub-genre in particular, ecocinema, is adopting this epistemological approach to further retrain its audience's perception and motivate them to act through ways that are more semiotic than other more traditional documentary approaches.

Bibliography

Armatage, Kay, et al., eds. "Studio D's Imagined Community." In *Gendering the Nation: Canadian Women's Cinema*. Toronto: University of Toronto Press, 1999. Print.

Aufderheide, Patricia. *Documentary Film: A Very Short Introduction*. New York: Oxford University Press, 2007. Print.

Costa, José Manuel. "Joris Ivens and the Documentary Project." In *Joris Ivens and the Documentary Context*, ed. Kees Bakker. Amsterdam: Amsterdam University Press, 1999. Print.

Enright, Robert. "Vérité, Vérité, All is Vérité." *BorderCrossings* 19, no. 3 (August 2000). Print.

Feldman, Seth. "Peace between Man and Machine." In *Documenting the Documentary: Close Readings of Documentary Film and Video*, ed. Barry Keith Grant and Jeannette Sloniowski. Detroit: Wayne State University Press, 2014. Print.

Ferreira, George, Ramirez Ricardo, and Allan Lauzon. "Influencing Government Decision Makers through Facilitative Communication via Community-Produced Videos: The Case of Remote Aboriginal Communities in Northwestern Ontario, Canada." *Guelph: Journal of Rural and Community Development* 4, no. 2 (2009). Print.

Gaines, Jane M. "Political Mimesis." In *Collecting Visible Evidence*. Minneapolis: University of Minnesota, 1999. Print.

Grierson, John. *Cinema Quarterly* 2.1, Edinburgh: G.D. Robinson, 1933. Print.

Hogenkamp, Bert. "The Radical Tradition in Documentary Film-Making, 1920s–50s." In *The Documentary Film Book*. London: British Film Institute, 2013. Print.

"Introduction to Fogo Island." *National Film Board of Canada*. Web. Accessed July 3, 2015. https://www.nfb.ca/film/introduction_to_fogo_island.

Jaffe, Ira. *Hollywood Hybrids: Mixing Genres in Contemporary Films*. Plymouth, UK: Rowman and Littlefield Publishers, Inc., 2008. Print.

James, Steve. "We Aren't Sorry for this Interruption." In *Screening Truth to Power: A Reader on Documentary Activism*. Montreal: Cinema Politica, 2014. Print.

Jones, D.B. *Movies and Memoranda: An Interpretive History of the National Film Board of Canada*. Ottawa: Canadian Film Institute, 1981. Print.

Kara, Keeling. "The Image and Common Sense." In *The Witch's Flight: The Cinematic, the Black Femme, and the Image of Common Sense*. Durham: Duke University Press, 2007. Print.

Malitsky, Joshua. *Post-Revolution Non-Fiction Film: Building the Soviet and Cuban Nations*. Bloomington, IN: Indiana University Press, 2013. Print.

Marchessault, Janine, ed. "Amateur Video and Challenge for Change." In *Mirror Machine: Video and Identity*. Toronto: YYZ Books, 1995. Print.

Michelson, Annette, ed. *Kino-Eye: The Writings of Dziga Vertov*. Berkeley, CA: The University of California Press, 1984. Print.

Nash, Terre. "Against the Grain: In 1974, When Kathleen Shannon Founded Studio D." *This Magazine*, 31, no. 6 (May–June, 1998). Print.

Nichols, Bill. "How Have Documentaries Addressed Social and Political Issues?" In *Introduction to Documentary*. Bloomington and Indianapolis: Indiana University Press, 2001a. Print.

———. "People as Victims or Agents." In *Introduction to Documentary*. Bloomington and Indianapolis: Indiana University Press, 2001. Print.

———. "The Voice of Documentary." In *New Challenges for Documentary* (2nd ed.). Manchester and New York: Manchester University Press, 2005. Print.

———. *Introduction to Documentary* (2nd ed.). Bloomington and Indianapolis: Indiana University Press, 2010. Print.

O'Flynn, Siobhan. "Documentary's Metamorphic Form: Webdoc, Interactive, Transmedia, Participatory and Beyond." *Studies in Documentary Film* 6, no. 2 (2012). Print.

Ostherr, Kirsten. "Animating Informatics: Scientific Discovery through Documentary Film." In *A Companion to Contemporary Documentary Film*. Chichester, West Sussex: John Wiley & Sons, Ltd., 2015. Print.

Pearson, Roberta, and Philip Simpson, eds. *Critical Definition of Film and Television Theory*. New York: Routledge, 2001. Print.

Plantinga, Carl. *Moving Viewers. American Film and the Spectator's Experience*. Los Angeles: University of California Press, 2009. Print.

Renov, Michael, ed. "Introduction: The Truth About Non-Fiction." In *Theorizing Documentary*. Los Angeles: The American Film Institute, 1993a. Print.

———, ed. "Towards a Poetics of Documentary." In *Theorizing Documentary*. Los Angeles: The American Film Institute, 1993b. Print.

Rosenthal, Alan. *Challenges for Documentary*. Berkeley: University of California Press, 1988. Print.

Royal Commission Report on the Status of Women. Ottawa: Government of Canada, 1970. Print.

Rusted, Brian. "The Politics of Distribution and Counterpublics." In *Challenge for Change: Activist Documentary at the National Film Board of Canada*, ed. Thomas Waugh, et al. Montreal and Kingston: McGill-Queen's University Press, 2010. Print.

Sherr Klein, Bonnie, and Dorothy Todd Hénaut. "In the Hands of Citizens: A Video Report." *Challenge for Change: Newsletter*, no. 4 (Spring–Summer, 1969). Print.

"The Power of Film." *Impact Guide*. New York and London: BRITDOC. Accessed September 4, 2016. http://impactguide.org/introduction/the-power-of-film.

Vanstone, Gail. *D is for Daring: The Women Behind the Scenes of Studio D*. Toronto: Sumach Press, 2007. Print.

Véronneau, Pierre. "The Cinema of Quebec." In *The Canadian Encyclopedia*. Historica Canada. Toronto. Web. Accessed June 8, 2015. https://www.thecanadianencyclopedia.ca/en/article/the-cinema-of-quebec.

Watson, Patrick. "Challenge for Change." *artscanada* 27, no. 2 (April, 1970). Print.

Williams, Linda. "Mirrors Without Memories: Truth, History and the New Documentary." In *The Documentary Film Reader: History, Theory, Criticism*, ed. Jonathan Kahana. New York: Oxford University Press, 2016. Print.

Winston, Brian. "The Tradition of the Victim in Griersonian Documentary." In *New Challenges for Documentary*. London: University of California Press, 1988. Print.

———. *Claiming the Real*. London: British Film Institute, 1995. Print.

Filmography

An Inconvenient Truth, Director, Davis Guggenheim. Lawrence Bender Productions, 2006.

If You Love This Planet, Director, Terre Nash. National Film Board of Canada, 1982.

Komsomol (aka *Songs of Heroes*), Director, Joris Ivens. Mezhrabpomfilm, 1932.

Kon Tiki, Director, Thor Heyerdahl. Artfilm, 1950.

Man with a Movie Camera, Director, Dziga Vertov. Moscow: VUFKU, 1929.

March of the Penguins, Director, Luc Jacquet. National Geographic Films (as National Geographic Feature Films), Bonne Pioche, Wild Bunch, Alliance de Production Cinematographique (APC), Buena Vista International Film Production France, Canal+, L'Institut Polare Français Paul-Émile Victor, 2005.

Misère au Borinage, Directors, Joris Ivens and Henri Storck. Social Films, 1933.

Nanook of the North, Director, Robert Flaherty. Les Frères Revillon, 1922.

New Earth, Director, Joris Ivens. Capi-Holland, 1933.

The 400 Million, Director, Joris Ivens. History Today, Inc. 1939.

The Newfoundland Project (*aka The Fogo Film Project*), Director, Colin Low. Montreal: National Film Board of Canada, 1967–1968.

The Spanish Earth, Director, Joris Ivens. Contemporary Historians, Inc., 1937.

The Thin Blue Line, Director, Errol Morris. Third Floor Productions, 1988.

The Things I Cannot Change, Director, Tanya Ballantyne. Montreal: National Film Board of Canada, 1967.

VTR St-Jacques, Director, Bonnie Sheer Klein. Montreal: National Film Board of Canada, 1967. Accessed September 7, 2016. https://www.nfb.ca/film/vtr_st_jacques.

Working Mothers, Director, Kathleen Shannon. Montreal: National Film Board of Canada, 1974–1975.

Yosemite: The Fate of Heaven, Director, Jon Else. No Production Company, 1989.

CHAPTER 4

Ecocinema and Semiotic Storytelling

4.1 Introduction

Much of the theory examined in the previous chapter addresses how the documentary film, its styles, forms, and modes, along with its accompanying technologies and exhibition platforms, can be used to enhance its ability to successfully influence and inform those charged with creating progressive social change. This chapter focuses on the theory behind one sub-genre in particular, and its semiotic approach to storytelling: ecocinema, a term first coined by film critic Scott MacDonald in his 2004 essay "Toward an Eco-Cinema" (MacDonald, "Toward an Eco-Cinema," 109). This specific approach is profiled as it augments the documentary film's ability to inform power and influence social change in a unique fashion. It also serves as a specific approach taken to produce a geomedia project I have created for the international environmental policymakers of the United Nations that will be examined in Chap. 7.

Ecocinema covers a lot of ground, literally, metaphorically, and even geospatially, so for the purpose of this chapter and to maintain focus on this book's thesis, this chapter will not be examining the fictional films that comprise this critical area. Instead, particular attention will be given to the environmental documentary—or as it is commonly known, the "eco-doc"—in its early forms and how new and emerging theory and practice are shaping its use as an instrument of global change in socio-political and geo-political policy and educational practice. A fair question to ask is why

M. Terry, *The Geo-Doc*, Palgrave Studies in Media and Environmental Communication,
https://doi.org/10.1007/978-3-030-32508-4_4

has the environmentally themed documentary been privileged with its own set of production theories; what is it about stories related to environmental issues that require special attention to move audiences to action more than other documentaries examining other social issues?

One of the answers often provided in ecocriticism is that ecocinema addresses environmental issues that are global in nature, and few other social issues are as comprehensively universal to the human experience as those of the environment. The eco-doc often depicts the *real* world outside the cinema in which the audience lives and thereby establishes a strong connection with the message on screen in the *reel* world. This invisible bridge between the world on screen and the world off screen successfully links the audience's mental, intellectual, and emotional engagements to the two worlds, virtually eliminating any non-relational reaction that may result from viewing any other kind of documentary. Belinda Smaill describes this relational process as an evolution of the traditional documentary film: "The eco-doc can be contextualized within a history of documentary filmmaking while also offering its own specific mode of address to the viewer" (Smaill, "Emotion, Augmentation, and Documentary Traditions," 104).

But how does ecocinema differ from environmental documentaries? First of all, ecocinema is a generic category for both fiction and non-fiction films that address environmental issues or themes. The environmental documentary can showcase a variety of environmental themes, and while often didactic in nature and frequently with a goal of moving its audiences to action, there can be environmental documentaries that do neither: the top ten waterfalls of the world, mountain-climbing expeditions, a tour of Caribbean beaches. These are subjects that, while being environmental in content, are not socio-political in context. A key clue to the activist intentions of ecocinema is its prefix "eco." *Eco* originates from the Greek word *oikos*, meaning a house and those who live within it. With this in mind, we see the process-relational importance of ecocinema to connect audiences with messages of their home and their place in it. A documentary on lunar studies might be intellectually interesting, but not emotionally engaging, since we do not relate to it in the same way as a documentary about earthly habitats. Documentaries that tell stories not only of our *home*, but our place *in it* have the potential to engage the viewer emotionally and compel them to action if the message is how we are acting as *homewreckers*. With the abundance of natural resources that have always been available to us, we have developed a complacency about our consumption of these commodities. Only a handful of times in the past have we recognized our harmful engagement with our global home environment (smog, air pollution, acid rain, the ozone hole), and each time we have acknowledged our direct contribution to these problems by way of industrial practice and capitalistic pursuits that often place environmental considerations behind revenue-generating ones. We frequently forget that our activity on this planet has a direct impact on the global ecology, and adaptation to our anthropogenic influence is rarely guaranteed.

The significance of understanding the difference between environmental documentaries and ecocinema documentaries is described from the perspective of audience influence and call to action by Paula Willoquet-Maricondi in her chapter "Shifting Paradigms: From Environmentalist Films to Ecocinema":

> An important distinction … between 'environmentalist' films and ecocinema is the latter's consciousness-raising and activist intentions, as well as responsibility to heighten awareness about contemporary issues and practices affecting planetary health. Ecocinema overtly strives to inspire personal and political action on the part of viewers, stimulating our thinking so as to bring about concrete changes in the choices we make, daily and in the long run, as individuals and as societies, locally and globally. (Willoquet-Maricondi, *Framing the World: Explorations in Ecocriticism and Film*, 45)

4.2 The Storytelling Modes of Ecocinema

In order to achieve these goals, ecocinematic documentaries employ a variety of modes and storytelling techniques to engage audiences not only emotionally, but activistically as well. It is not enough to merely have one's conscious raised, but one must be compelled to use this new information to actively bring about progressive change. One of the most popular techniques is the use of fear. It is relatively easy to invoke fear in ecocinema—almost all ecocinematic fiction does this: (*The Day After Tomorrow* (2004), *2012* (2009), *Quintet* (1979), *Soylent Green* (1973), *Geostorm* (2017), *Waterworld* (1995), *Snowpiercer* (2014), to name just a few); however, in non-fiction films, the eco-doc has the added responsibility of measuring its dosage of fear appeal and to temper it with suggested courses of action for its audiences to take.

The danger with trying to engage an audience through the emotion of fear is the possibility of disengagement. Fear could induce the viewer to "look away" and detach from the message the film aims to convey. Fear in documentary ecocinema is often identified as fear-mongering which has an even more profound negative impact on audiences as the message can be perceived as untrustworthy and propagandistically manipulative. However, when used responsibly and in the right dosage, fear can be an effective mobilizer of audience action, as Mei-Fang Chen concludes in a paper, testing audiences of environmental documentaries, "Impact of Fear Appeals on Pro-environmental Behavior and Crucial Determinants":

> Fear is a powerful, innate emotional response to a perceived threat or dangerous event and a strong motivator for people to change their behavior to avert a potentially negative outcome. Negative emotions can be used to induce specific moods or emotions to persuade people to change their thoughts or behaviors, particularly if certain behaviors have dangerous or adverse consequences. (Chen, 75)

Specifically, Chen finds that low and moderate fear appeals in documentary films about climate change are effective in encouraging pro-environmental behaviour, but only if "feasible recommended actions" (Chen, 86) accompany the film's message. This works when the level of fear appeal is presented "in a factual and less personal message without exaggerated negative consequences" (Chen, 87). When an eco-doc engages in high fear appeals, they "induce a threat that is too severe, resulting in full absorption in defensive mechanisms rather than attention to or complying with a proposed solution" (Chen, 75, 76).

Emotional engagement is a requirement to induce audiences to action, but it is equally important to connect audiences' view of the environmental world on screen with their view of the environmental world off screen to understand that the issues portrayed in the eco-doc have a direct impact on them, where they live, and their futures.

One of the best documentaries to present the intrinsic connection between humans and their natural habitats is Godfrey Reggio's 1982 film *Koyaanisqatsi*, made in the poetic mode of documentary film. The title refers to a Hopi Indian word meaning "life out of balance," and the film presents its argument for how this is true to human life as well as planetary life through a series of striking images that often blur the line between human activity and natural phenomenon. Rocket exhausts and ash look like storm clouds and snow; time-lapse scenes of traffic look like colourful rivers; and the chaotic movement of people at Grand Central

Station looks like insects swarming around their colony. This blending of humans with their environment disrupts our traditional viewing experience of these images. The time distortion techniques of slow motion and time-lapse, the lack of conventional editing, the absence of narration or subject interviews, all combine to impose an altogether different viewing experience engineered to "retrain our perception" as Scott MacDonald demands of ecocinema as a means of engaging the audience in a more profound manner (MacDonald, "The Ecocinema Experience," 45). Gary Matthew Varner's paper "*Koyaanisqatsi* and the Posthuman Aesthetics of a Mechanical Stare" explains how Reggio uses unconventional methods to present natural environments juxtaposed against the mechanically reproduced environments:

> The unnatural vision produced by time-lapse remains consistent regardless of what fixes the camera's attention. In the film's second half, foot and car traffic are presented to viewers no differently than the movement of ocean or clouds. Koyaanisqatsi's form is not interested in a rift between the natural and the cultural. (Varner, "*Koyaanisqatsi* and the Posthuman Aesthetics of a Mechanical Stare," 2017)

In Graham Cairn's book, *The Architecture of the Screen: Essays in Cinematographic Space*, he describes how these unconventional images force the audience to see the familiar in a different way:

> For all its use of juxtaposition … Reggio's cinematographic games go way beyond simple techniques of fragmentary and multiple edition; they involve a very particular and intelligent approach to pictorial composition … Playing with our perception of what appears on screen, through the strict control of the subject and point-of-view relationship, Reggio reveals just how important composition is. (Cairns, 78)

There are, in fact, several modes of address currently being used by ecocinematic documentary filmmakers, and, no doubt, many more to come as they are conceived, all designed to cause audiences to see the familiar in a different way so they can eschew complacency and become self-motivated to act. In Claire Ahn's 2018 thesis *Visual Rhetoric in Environmental Documentaries*, she identifies as many as six modes: ***apocalyptic***, ***jeremiad***, ***hopeful***, ***environmental nostalgia***, ***sublime***, and ***environmental melodrama*** (Ahn, iii). Films made in the *apocalyptic* mode provide evidence of impending natural catastrophe and often project their impacts on humans. These images are intentionally dark and horrific, so the viewer will be more inclined to take action to prevent the depicted eventuality. One of the best examples of this mode is the television series *Life After People*, which this chapter will examine in detail as an example of a related mode, ***anticipatory***. The ***jeremiad*** mode takes dead aim at the prevailing cause of our environmental woes, us. Anthropogenic themes are common as the filmmakers attempt to reveal the specific careless actions usually committed in capitalistic pursuits and the effects of these actions on our global ecology. The goal of this mode is to enlighten the viewers as to the cause of the environmental crisis and then compel them to take action to stop this behaviour. Josh Fox's *Gasland* (2010) and Louie Psihoyos' *Racing Extinction* (2015) are primes examples of the jeremiad mode. The *hopeful* mode steers clear of the ugliness of apocalyptic imagery, choosing instead to portray the beauty of nature and the possible solutions to reverse our destructive directions with an optimistic tone. My own films *The Antarctica Challenge: A Global Warning* (2009) and *The Polar Explorer* (2010) have been categorized by scholars, film critics, and film festival programmers as being made in this mode. The ***environmental nostalgia*** mode tells stories that harken back to the good, old days before capitalistic endeavours thoughtlessly polluted our air, water, and land, deforested our landscapes, and levelled mountains in the pursuit of profit. By seeing what our environment looked like before compared to how the same environment looks today, a powerful contrast is instantly visible, evoking a strong emotional response from audiences and motivating them to take action. Jeff Orlowski's films *Chasing Ice* (2012) and *Chasing Coral* (2017) exemplify this mode with stunning before-and-after time-lapse photography of beautiful and healthy polar ice and sea coral turning into destroyed fragments and faded shades of their former selves. The ***sublime*** mode endeavours to draw a personal connection between the audience and what they see on the screen, usually a spectacular scenic vista or an animal. These images are presented against a background of an environmental issue that threatens these landscapes and creatures. The personal connection made by combining beauty with danger is designed to compel audience member to act to save and protect them. Gabriela Cowperthwaite's film *Blackfish* (2013), discussed previously in this book, and Luc Jacquet's *March of the Penguins* (2005), examined later in this chapter, are good examples of the effectiveness of the sublime mode. ***Environmental melodrama*** is perhaps the most emotionally manipulative mode in ecocinema as it plays to the highly relational aspects of how environmental issues directly impact on people's lives. Steven Schwarze, an associate professor in the Department of Communication Studies at the University of Montana, sees melodrama as a valuable strategy in audience engagement:

> (Melodrama) can transform ambiguous and unrecognized environmental conditions into public problems; it can call attention to how distorted notions of the public interest conceal environmental degradation; and, it can overcome public indifference to environmental problems by amplifying their moral and emotional dimensions. To the extent that melodrama serves these purposes, it presents itself as an enticing rhetorical strategy for environmental advocates. (Schwarze, 239–240)

The environmental melodrama mode has a long history in documentary, going all the way to Pare Lorentz's *The Plow that Broke the Plains* (1936) examined in Chap. 2. It has survived all these years because it is so effective. A more recent example of this mode is Jon Shenk's moving look at rising sea levels displacing the 400,000 residents of the Maldive Islands, *The Island President* (2012).

What all these modes have in common is an aim to change the way audiences perceive the content of this kind of documentary story. As ecocinema addresses global environmental issues that impact all of us, a universal technique that reaches a global audience needs to be introduced the way the images and related narratives are presented to the audience. The traditional ways merely inform and often emotionally engage, but in order to achieve audience action, new ways of reaching the audience and compelling them to act after they leave the cinema become the aim of ecocinematic documentaries.

In line with this strategy, ecophilosophical theorists such as Felix Guattari advocate that a retraining of perception is required in all sociopolitical areas, including the way environmental documentaries are made and used as communications tools to inform power. Guattari argues that this is necessary to provide a fuller understanding of the global threat of many environmental issues that we ourselves are causing, but may not be aware of. In his essay "Chaosmosis: An Ethico-aesthetic Paradigm," Guattari argues for the need to change the way we perceive environmental issues to understand and act on more effective levels:

> Our survival on this planet is not only threatened by environmental damage but by a degeneration in the fabric of social solidarity and in the modes of

> psychical life, which must literally be reinvented. The refoundation of politics will have to pass through the aesthetic and analytical dimensions implied in the three ecologies—the environment, the socius, and the psyche. We cannot conceive of solutions to the poisoning of the atmosphere and to global warming due to the greenhouse effect, or to the problem of population control, without a mutation of mentality, without promoting a new art of living in society. (Guattari, *Chaosmosis*, 20)

Guattari's "mutation of mentality" is required, he argues, since environmental issues require a new way of being seen and conceived to have mankind change its relationship with it and to change the mistakes it has already made to it. Therefore, the traditional methods of audience engagement used in documentary film with environmental themes need to be similarly adjusted. The use of film to reveal the world in terms of environmental themes goes back to the very origins of film itself. Some of the very first documentary feature-length films were environmentally themed, aimed at bringing unknown parts of the world to the rest of the world. In particular, the polar regions seemed to hold special interest. Films like *South* (1919), *Romance of the Far Fur Country* (1920), *Nanook of the North* (1922), and *A Sixth Part of the World* (1926) were all shot in whole or in part in either the Arctic or Antarctica. With the exception of *South*, Sir Ernest Shackleton's film log of his exploratory voyage to Antarctica, the other films showcased ethnographic profiles of people (Inuit and Sami, primarily) native to these frozen climes and how their communities were impacted by national, capitalistic and technological colonialism.

These early forays into documenting the natural world relied less on story and more on observation. Curiously, the genre is coming full circle as new theory suggests a return to this expositional observance, but with semiotic, implicative narratives to establish Guattari's "mutation of mentality," the primary goal of ecocinema, as Scott Macdonald echoes with his call for a "retraining of perception" in his essay "The Ecocinema Experience":

> The fundamental job of eco-cinema [sic] … is a retraining of perception, as a way of offering an alternative to conventional media spectatorship, like a garden within the machine of modern life. (MacDonald, "The Ecocinema Experience," 45)

The garden Macdonald refers to invokes images of environmental content consistent with most ecocinematic films, but it also serves as a metaphor of a thriving ecology that is constantly growing and being maintained by filmmaker-gardeners. This evolving ecology has seen a significant change made to the garden's very environment: digital imaging. In the past, celluloid was the only way of capturing reality and sharing those images with viewers throughout the world for the documentary film. While this medium was impressive, it took a mere one hundred years to replace it almost completely with a format that is not only less vulnerable to deterioration, but able to reach many more people in a much more immediate fashion.

As a sub-genre of documentary film, ecocinema defines itself in epistemological terms with a raison d'être to educate, inform, and then move audiences to action that yields extensive and beneficial change for both the planet and its inhabitants. Paula Willoquet-Maricondi provides this definition in her book *Framing the World: Explorations in Ecocriticism and Film*: Ecocinema has "consciousness-raising and activist intentions, as well as responsibility to heighten awareness about contemporary issues and practices affecting planetary health" (Willoquet-Maricondi, 2). However, consciousness and awareness-raising are not simply achieved by providing compelling information. In ecocinema, new ways of engaging the audience both mentally and emotionally are being developed and practised; the audience's thinking is tapped in new ways to assist the viewer in deciding to act on their own, and not only through explicit exposition.

Many of these new storytelling techniques have emerged, thanks to the digital tools and related affordances not previously available to the documentary filmmaker. Ecocinema, in terms of documentary form intended to influence audiences to action, is a cinematic method of presenting environmental issues in a didactic and balanced manner incorporating explicit *and* implicit narratives afforded by digital technologies, theories, and practices. Ecocinematic films can be both familiar as documentary films and unconventional in narrative, combining the two in an altogether new way to show humans the world they have so adversely impacted and to encourage them to act to create progressive environmental and behavioural change on a global basis.

4.3 Performing the Database Documentary

Along with this new technology came new ways of recording and manipulating images, both in terms of enhancement and distortion. These new digital tools have afforded the documentary filmmaker creative new methods of storytelling, methods best explored by ecocinema. DJ Spooky, aka Paul D. Miller, is a digital artist who, like his cinematic ancestors, created an ecocinema project (which he refers to as a "product") showcasing a polar region, Antarctica, in his database documentary *Terra Nova: Sinfonia Antarctica* (2008). His experimental piece is not a traditional documentary viewing experience. It is a film performed by the filmmaker; as images of Antarctica play out on two big screens, DJ Spooky himself mixes music live along with live performances provided by a pianist, violinist, and cellist (see Fig. 4.1). At times, the filmmaker will provide live commentary, as opposed to descriptive narration, as he explains why he made the film while it is playing. This unusual artistic approach pays homage to the early days of James Freer's "animated lectures" and silent film's "musical accompaniment," not to mention a database documentary nod to Walter Ruttman's 1927 showcase of an urban environment, *Berlin: Symphony of a Great City*, but while reminiscent of these pioneering days of the

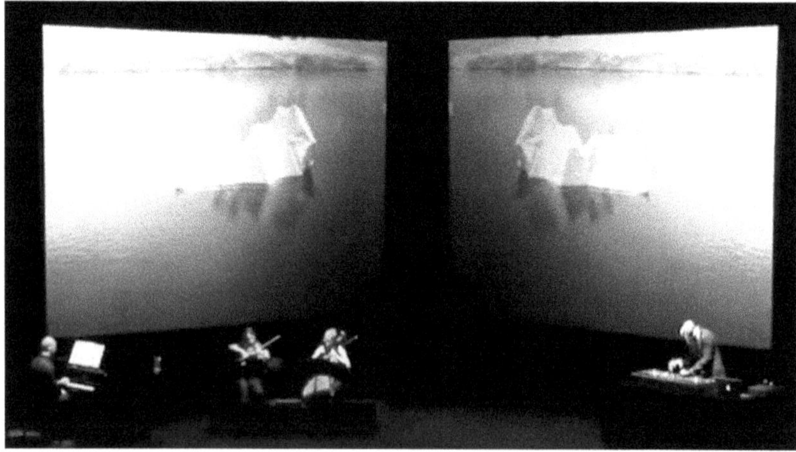

Fig. 4.1 Scene from *Terra Nova: Sinfonia Antarctica*. New York: No production company, 2008. https://www.youtube.com/watch?v=zJ7Y7GzxWZM. Frame of scene occurs at 00:01:45

documentary, Spooky's modern use of these audio techniques has an altogether different intent.

> I think that people need to *hear* Antarctica because it is at the edge of the world ... New York is probably one of the most mediated places on earth. If I have a conversation at a café, someone will put it on a blog. If I walk down the street, someone will put photos of it on Flickr ... Antarctica represents a place mediated by science. (*Italics are mine*) (Shembel, 345)

Spooky's goal appears to be introducing a relatively unknown part of the world (on ancient maps, Antarctica was designated as "Terra Nova" (New Land) and, alternately, "Terra Incognita" (Unknown Land)) to the people of the world living outside the scientific community. The disparate visual content includes the traditional tropes of icebergs and penguins, but also anthropogenic images of early exploration, scientific research, flags of the signatory nations of the Antarctica Treaty, and cartography which all speak to man's attempt to understand, colonialize, and perhaps even tame this last of the world's wild frontiers.

Non-traditional narratives play a big role in ecocinema projects such as this. Instead of guiding the audience through a story with subject interviews and narration, new and innovative ways of informing documentary audiences are being tried. Spooky's focus on audio—live voice, effects, and music—is one way ecocinema is retaining the perception of its audiences. Screening images without apparent context to any recognizable storyline is another way. With this method, coupled with live audio cues, audiences are challenged to construct their own story, to be the film's editor as it were and assimilate the database of sounds and images presented to formulate their own story and conclusions. In so doing, they are uniquely motivated to act taking direction from themselves instead of from the film's director, as is often the intent of an activist documentary.

Spooky has said this project is continually in development—a "living" documentary whose ecocinematic gardener—to continue Macdonald's metaphor—lovingly maintains and nurtures so that its ever-changing flowers may attract and motivate new audiences indefinitely. As we will explore in forthcoming chapters, the multilinear documentary and the database documentary in today's digital domain are popular and successful narrative approaches for the ecocinema filmmaker.

4.4 Semiotic Storytelling

I make a distinction between the activist documentary and the social issue documentary: not all social issue documentaries are activist documentaries, but most activist documentaries are social issue documentaries. The difference here is in relation to audience. The activist documentarian will create a film in which the narrative often has a specific agenda to motivate its audience to action. The content is frequently one-sided in its approach to the social issue depicted, lacking a balanced journalistic presentation of the facts that allows the audience to decide for themselves. Propagandistic in nature, the activist documentary frequently succeeds in its approach to influencing and motivating its audiences to take action. Ecocinema theory embraces the less heavy-handed social issue method of documentary filmmaking and often employs both journalistic balance and implicit narratives afforded by semiotic storytelling to motivate its audiences in less explicit, more honest, and less forcefully intentional ways.

In their book *Ecocinema Theory and Practice*, Stephen Rust et al. claim that the essential purpose of ecocinema "should not be to impose a political program—much less pre-defined aesthetic practices—but to help create public spaces for debate and ethical argument over the claims of the environment for a place in political life" (Rust et al., 295). In so doing, ecocinema establishes an ethical countenance in its content and presentation before defining the theoretical approaches it takes to inform and influence its audiences. One of the most attributed philosophical reconceptions of the representations of environmental themes is the way audiences perceive the content now delivered in an unconventional form. Adrian Ivakhiv describes this approach to ecocinema filmmaking as a "machine that moves us along vectors that are affective, narrative and semiotic in nature and discloses worlds in which humanity, animality and territory are brought into relationship with each other" (Ivakhiv, "An Ecophilosophy of the Moving Image," 87).

This ecophilosophical interpretation of the documentary film owes a lot to Gilles Deleuze's theories on the movement-image and the time-image he describes in his *Cinema 1* and *Cinema 2* books, respectively. In particular, the affective nature of film begins with perception that causes affect—an emotional reaction—in the viewer, leading them to action. In documentary terms, Siobhan O'Flynn defines "experience design" as addressing form and content, including affect, as the "physical experience of the user beyond the intellectual comprehension of information, emotional impact and includes human-to-human interaction and calls to action" (O'Flynn, 74). Deleuze describes these narrative elements as the perception-image, the affection-image, and the action-image. In general terms, the perception-image is that which the audience sees first when viewing a film, frequently referred to in cinematic terms as the establishing shot; the affection-image is how those perceptions affect the viewer, often the close-up; and the action-image is the resulting action either taken on screen (the two-shot) or, as I argue in ecocinema, by the audience. Deleuze identifies the documentary as one of the primary film genre's most effective in its use of the action-image.

Ecocinema's interpretation of Deleuze's movement-image theory is to underscore explicit narratives with implicit narratives via semiotic techniques. In Deleuze's *Cinema 2* book, he explains that the time-image in film imbues meaning through speed, and when we distort that speed we change our perception, breaking the cliché, as Keeling and Plantinga assert, to cause the viewer to think actively as opposed to view passively. Much of ecocinema documentary employs this method of duration distortion and is sometimes categorized as "slow cinema," a style in which the scenes are uncomfortably long to the point of forcing the audience to question why.

In the film *The Antarctica Challenge: A Global Warning*, an eco-doc I made in 2009, this technique was used in the first scene that presented a glacier. The long pan without narration took forty seconds, a lengthy period of time in film to view a scene in which nothing happens. This slow movement was intended to reflect and underscore the slow movement of glaciers. Historically, the melting and the subsequent movement of these frozen rivers are very slow, and this scene is equally slow to highlight this fact. Once this has been established to the audience, the next glacier scene is antithetically fast. A zoom from space using satellite photography quickly lands on the Pine Island Glacier on Antarctica's west coast, the largest ice drainage basin in the world. Once the glacier location is established, time-lapse photography shows the uncharacteristic rapid movement of the glacier as it moves freshwater ice from the land to the sea. The semiotic hyper-speed time-image is intended to present an unfamiliar perception of a glacier to affect the audience to a degree that enables them to see the urgency in natural action and in political reaction.

Made in the *hopeful* mode of ecocinema, this documentary resonated with scholars (Claire Ahn 2018), film critics (Christopher Heard, 2010), and film festival programmers (Will Nugent, International Film Festival Ireland, 2010) for being made in this mode. The journalistic approach of data presentation and interviews with scientists, as opposed to activists, appealed as well to the international environmental policymakers of the United Nations, which this book will examine in more detail in Chap. 7. Evidently, an eco-doc made in this mode, with a journalistic approach and incorporating semiotic storytelling techniques, presents an effective structure for ecocinema projects aimed at influencing the changemaker.

Ecocinema theorists see great value and impact on audience engagement through these Deleuzian, semiotic practices. A frequent collaborator of Deleuze, Felix Guattari wrote about this approach in his book *The Three Ecologies* (2000), often considered the first to identify the effectiveness of semiotic narratives in environmental documentaries; Ivakhiv's thesis on the subject is called *Ecologies of the Moving Image: Cinema, Affect, Nature*; and Alexa Weik von Mossner edited a volume of essays curating a collective of scholarly perspectives on the subject entitled *Moving Environments: Affect, Emotion, Ecology, and Film* (2014).

Guattari's three ecologies are social ecology, mental ecology, and environmental ecology. They represent a definition of ecology that extends beyond conventional environmental terms to include human subjectivity and, as such, provides a useful framework to examine how audiences perceive, are affected by, and act upon the documentaries of ecocinema.

Related to Deleuze's theory of perception as a function of the movement-image in film viewing, Guattari stresses the need to look beyond the capitalistic imperatives that rule most people's everyday lives. Instead of perceiving a story in terms that impact the viewer specifically and individually—will I have to move if sea levels rise, for example—Guattari argues that we need to "reevaluate the purpose of work and human activities according to different criteria than those of profit and yield" (Guattari, *The Three Ecologies*, 57), thus providing the retraining of perception Macdonald advocates. Once this is achieved, the imperatives of mental ecology—"the appropriate mobilization of individuals and social segments as a whole" (Guattari, *The Three Ecologies*, 57)—are possible, the ultimate goal of ecocinema in general and activist documentary in particular.

Guatarri's social ecology has a symbiotic relationship with this. Its principle "concerns the development of affective and pragmatic cathexis in human groups of different sizes" (Guatarri, *The Three Ecologies*, 60). Once the perception of the viewer is retrained to see ecocinema's messages as pertaining to groups of people rather than individuals, global rather than local themes, the concentration of mental energy of a specific idea affects the viewer in a more profound way that motivates action. This theme is reinforced by other ecocinema scholars, such as Jason W. Moore who, after being questioned by one of his students, came to understand the Anthropocene (the current geological epoch which is defined by mankind's direct influence on the global environment) as being more accurately viewed as the *Capitalocene*, since the majority of our human activity that has impacted our planet is based on Guattari's theory of "profit and yield" (Moore, xi).

A further reevaluation of our perception is required in the view and practice of environmental ecology, argues Guatarri. He promotes a restructuring of the current social mobilization of "small groups" of environmental activists

> who sometimes deliberately refuse any large-scale political involvement. Ecology must stop being associated with the image of a small, nature-loving minority or with qualified specialists. Ecology in my sense questions the whole of subjectivity. (Guatarri, *The Three Ecologies*, 52)

The subjectivity to which he refers is the objectivity of environmental ecology, that it be recognized as a global concern that affects all people, everywhere. This, in turn, helps define ecocinema as a significant communications tool to inform, influence and activate progressive social change for the world. Mental and social ecologies, as he defines, become valuable theories in ecocinema storytelling to achieve the goal of environmental ecology and sustainability for all.

Guatarri makes the point that is often made that human beings are part *of* the earth, not merely living *on* the earth and as such have a responsibility to be aware of and sensitive to our impact on it. Developing a social and mental ecology yields a moral consideration that helps sustain an ecology for the earth's environment. The interrelated nature of social, mental, and environmental ecologies is more transversal than universal in its application, making this ecosophy a foundation for ecocinema documentary projects. The filmmaker takes on the traditional and proven methods of social issue documentary filmmaking when developing narratives that affect and engage audiences mentally and socially, but when environmental themes are added, the creative ecology that results blossoms new interconnected methods of documentary storytelling, producing a stronger motivation for audience action.

Remediating the genre in this manner enhances audience perception, affect, and action. By providing an entirely new environment for viewing an ecocinematic documentary, as DJ Spooky conceived with his live audio performance and database documentary video montage, the semiotic messages of the media fragments are rationalized and conceptualized by the audience mentally and applied to action socially. Unlike an activist documentary that baldly instructs its audience to act, ecocinema provides as much data as possible in an introjective manner that allows the audience to derive the importance to act on its own, thus yielding a more self-conscious, self-motivated activist.

Guatarri's ecosophy plays an important role in Adrian J. Ivakhiv's detailed analysis of ecocinema and ecocriticism, *Ecologies of the Moving Image: Cinema, Affect, Nature*. In this book, he supports Guatarri's theory of inter-relational activity between perception, affect, and action using Deleuze's movement-image as a framework to analyse the particular structure of this new storytelling and audience-engaging process of ecocinematic projects. In his introduction, he argues that cinema is "world-making" in the same terms first identified by Martin Heidegger as poiesis (bringing into existence that which did not exist before) in his essay "The Origin of the Work of Art,"[1] and by Stanley Cavell's seminal work on the perception of cinema on audiences, *The Worlds Viewed*.[2]

This inter-related framework is a model Ivakhiv refers to as *process-relational*:

> It is a model that understands the world, and cinema, to be made up not primarily of objects, substances, structures, or representations, but rather of relational processes, encounters, or events. As we watch a movie, we are drawn into a certain experience, a relational experience involving us with the world of the film. In turn, the film-viewing experience changes, however slightly, our own experience of the world outside the film. (Ivakhiv, *Ecologies of the Moving Image*, 12)

This process-relational model is Ivakhiv's retraining of perception for ecocinematic documentary films. Through it, he argues, the two worlds—

[1] Heidegger, Martin. "The Origin of the Work of Art." *Off the Beaten Track*, Julian Young and Kenneth Haynes, editors and translators. Cambridge: Cambridge University Press, 2002.

[2] Cavell, Stanley. *The Worlds Viewed: Reflections on the Ontology of Film* (Enlarged Edition). Cambridge, MA: Harvard University Press, 1979.

on screen and outside the cinema—converge, providing the audience with a more profound engagement with the eco-doc and the actual environment depicted in the film:

> Films… can move us toward a perception of the world in which sociality (or the anthropomorphic), materiality (or geomorphic), and the interperceptual realm from which the two emerge are richer, in our perception, than when we started. This goes against the claims of those who have argued that technological mediation is more a part of the world's ecological problem than of its solution. (Ivakhiv, *Ecologies of the Moving Image*, 12)

In Ivakhiv's view, cinema, affect, and nature work together in ecocinema much the same way perception, affect, and action work in Deleuze's theory of audience engagement through the movement-image. Ivakhiv's cinema is a Heideggerian or Cavellian world which audiences willingly enter by devoting its perception to what is on screen and ignoring the other world, surrounding them in darkness; the audience is then affected by what they perceive, often emotionally, but also mentally, intellectually, and socially in ecocinema, as new information about their world outside the cinema is presented; finally, this affect moves an audience to action to improve its world or by raising an awareness that makes them behave and interact better within the natural world.

Another semiotic technique used by ecocinema documentary filmmakers according to Ivakhiv is the representation of "animal by analogy" (Ivakhiv, *Ecologies of the Moving Image*, 223). By presenting animals in the wild with storylines that parallel human behaviour, audiences relate better to the characters and immerse themselves in the presented environment more significantly. They are more apt to be affected by environmental issues seen on the screen when the anthropomorphized characters seem to mirror themselves so closely.

Ivakhiv examines the use of this technique in the phenomenally successful eco-doc, *March of the Penguins* (2005). The family values depicted by the Antarctic penguin colonies in the film moved many people, especially during the scenes of extreme sacrifice made by the parents in feeding and protecting their young during ferocious storms. The anthropomorphized penguins and their human-like lifestyles and values were championed by Christian conservatives and conservationists alike. The environmental issues presented subordinate to the penguin storyline provoked media attention on the subject of global warming, often with references to *March of the Penguins.*

By reaching out to audiences with this technique, *March of the Penguins* not only laid claim to being the second-highest grossing documentary of all time,[3] but succeeded in raising the level of awareness of the environmental issue of global warning by means that affected more people than other environmental documentaries that addressed the issue directly.

But simply raising awareness is not enough, and new methods of engaging the audience emotionally and affecting them to the point of moving them to action are the focus of new theories of practice in ecocinematic documentary films. In her book, *Moving Environments: Affect, Emotion, Ecology, and Film*, Alexa Weik von Mossner explores these methods through a collection of essays written by leading ecocritical scholars. She describes the entries in her anthology as “an intriguing exploration of the ways in which films engage their viewers’ ‘real’ and pre-existing emotion systems, relying not only on storytelling, but also on various filmic techniques such as lighting, editing, and music” (Weik von Mossner, 7).

One of the more intriguing and popular methods explored is the use of irony, an immigrant of fiction filmmaking now taking up residence in ecocinematic documentaries. In her contribution, “Irony and Contemporary Ecocinema: Theorizing a New Affective Paradigm,” Nicole Seymour argues that irony is a new tool for the ecocinematic filmmaker to draw attention to environmental issues that may be overlooked or not considered completely.

As a semiotic method, irony can help assist audiences in recognizing the bigger picture of environmental issues without hand-holding them through the process in an overly didactic and sometimes alienating way. In *The Antarctica Challenge: A Global Warning*, I filmed a scene of “teenage” penguins still being fed by their mothers instead of hunting for krill on their own. It was suggested that these young penguins are deciding to stay at home longer than previous generations did due to the “bleak prospects” of survival on their own in the outside world. The ironic humour suggested here is that the film’s human audiences are experiencing the same generational change in their own families. In question periods at film festivals, I was often asked what audience members could do to help increase the krill populations so the future for a new generation of penguins

[3] Box Office Mojo. Accessed January 1, 2018. http://www.boxofficemojo.com/genres/chart/?id=documentary.htm.

would not be so bleak. Here we see a direct desire to act resulting from the emotional affect caused by using irony in this way.

As discussed earlier in this chapter, *March of the Penguins* became a beacon of scientific evidence for conservative Christians who wished to prove that heterosexual family values represent the preferred lifestyle for all of God's creatures. Seymour suggests an ironic alternative film:

> we might imagine an eco-film produced from a campy, queer perspective—ironizing how nature is selectively valued for the ways in which it can advance human agendas, including homophobia. (Seymour, 65)

She adds that irony can also serve a more direct purpose for ecocinema by applying traditional tropes used in nature documentaries to unconventional environments. She suggests employing the reverential depiction of majestic mountains to inner-city environments. She correctly states that such a cinematic treatment would be "jarring, maybe even galvanizing" (Seymour, 65) as ecocinema ventures more into unconventional environments to explore and reveal the global issues that are impacting more and more on mankind in its own protected urban landscapes previously thought to be impenetrable to the ravages of nature: "as ecocinema increasingly expands its definitions of nature and environment to include less-than-ideal spaces and entities, irony may prove to be a crucial mode of operation" (Seymour, 65–66).

As crucial as it might be, the use of irony is not without its drawbacks in ecocinema rhetoric. As an effective tool, it can also be used by the propagandist for goals that do not necessarily fall in line with the progressive social and global change for which ecocinematic documentary filmmakers generally strive. In his essay, "The Post-ecologist Condition: Irony as Symptom and Cure," Bronislaw Szerszynski argues the flaw in the use of irony by environmentalists:

> the modes of irony typically deployed by environmental groups take a 'corrective' form, in that they draw attention to the gap between appearance and reality, or between stated intentions and behaviour, in order to try to overcome it—for example, by forcing corporations to act in conformity with their stated pro-environmental objectives. (Szerszynski, 337)

One of the dangers in this corrective form of irony involves the "privileged observer," who we call the filmmaker. Szerszynski rightly points out that the filmmaker "can *themselves* be accused of self-deception and bad faith"—ironically, the very things they accuse the anti-environmental capitalists they aim to convert—by being either not fully informed on the issue being represented or by having an agenda that deliberately omits reasonable data and arguments to the contrary. It is in this regard the activist documentary is exposed for its propagandistic tendencies. What works best, according to Szerszynski, is an "irony-as-world-relation," an irony that embraces even the observer/filmmaker, the one who identifies the irony. This echoes the journalistic approach I advocate in the creation of the social issue documentary and, in particular, the ecocinematic documentary. By presenting all sides of an issue in a documentary, as is the tenet of responsible journalism, the audience's decision to act is self-motivated, rather than directed. This yields a more powerful advocate for social change than the relatively minimally informed viewer who is acting on the word of the activist filmmaker. Irony, in this sense, is all inclusive, and the audience can therefore trust the observer/filmmaker to a greater degree.

This self-awareness proposed by Szerszynski is one of the strongest features of irony as a filmmaking tool for ecocinema for Seymour who sees the value in irony's self-reflexive and self-critical potential. Used to great extent in humanities studies and often referred to as the double gaze, this non-exclusive examination yields more data and knowledge, but it too comes at a price: "for all its potential to enable effective, non-elitist political action, irony can also make for inaction: for smug armchair environmentalism or even slacker apathy," Seymour warns. We take for granted that environmental documentaries are made with the goal of improving the planet, but do we question why it fails when it does? Irony assists in this self-reflexivity that can only help improve ecocinema's communicative properties and subsequent success with affect and action.

A final examination of a theory unique to ecocinema is one developed by Juan Francisco Salazar, a Chilean documentary filmmaker and theorist who advocates an additional function to Michael Renov's previously examined four: anticipatory. In his essay, "Anticipatory Modes of Futuring Planetary Change in Documentary Film," Salazar proposes an altogether new purpose for the documentary film in general, and for ecocinema in particular: to "act as a modality to render an anticipatory futuring of socio-ecological change" (Salazar, 44).

He refers to Tony Bennett's 1991 paper "The Shaping of Things to Come: Expo 88" which chronicles how visitors to the event held in Brisbane, Australia, encountered exhibits and technologies of the "future."

He explains that "World Expos project the future in the form of a task to be performed in the present" (Salazar, 44). By engaging with these future technologies, they make them concrete in the present, providing the guests with an experience that allows them to create "an anticipatory futuring of the self." In today's social issue documentary film and, in particular, those with environmental themes, Salazar argues that the same engagement is being introduced and sought by the filmmaker from their guests, the viewers. By seeing what the future may hold, the audience is asked to "future" themselves in the present to participate actively in avoiding environmental calamity, a form of cinematic crisis management.

In addition to Michael Renov's four functional modes of documentary—persuading, analysing, recording, and expressing—Salazar introduces a new one: *anticipating*. This mode promises the events depicted in the film are not just possible, but probable, requiring the audience to act in the present to prevent that which they experience in the film's future. As we have seen in other approaches explored in this chapter, this new mode proposed by Salazar is intended to engage and move audiences indirectly, a semiotic method to complement and underscore the message of the explicit narratives of ecocinematic documentary films. This anticipatory mode provides a new affect for audiences, one of desire to act. By witnessing what the future will hold by laying out a scientific data set that projects future actualities, such as in the History Channel's series *Life After People*, produced by Flight 33 Productions, the viewer is once again self-motivated to act today to prevent a calamity to come or to maintain current action to ensure an environmentally safe future for themselves.

Like Renov, Salazar acknowledges documentary's capacity for rendering desire in an audience—in this case, a desire to live in a utopian world that the film promises or a desire to avoid living in a dystopian world the film promises. Salazar builds on this, adding intent to this effect, arguing that together they yield stronger audience action:

> documentary cinema can potentially transcend the mode of desire, to encompass a modality of intent: to promote social change and induce a sense that a deep socio-ecological and economic transformation is needed to confront the uncertainties posed by a liquid future … I think documentary cinemas can play a role in instantiating the future by rendering it present and giving it a concrete form, thus permitting viewers to engage with anticipatory modes of futuring of the planet. (Salazar, 45)

By concretizing the future in the present, it becomes real, just like the exhibits at Expo 88, and a perceived reality of the present built on a convincing depiction of the future accomplishes something rarely considered in a medium that is commonly known as a technology that captures the past.

4.5 Conclusion

This chapter has examined several new and experimental theories employed by the makers of ecocinema, most of them semiotic in nature and used in conjunction with pre-existing theories of documentary modes and functions, but the question remains how effective these filmmaking approaches are in creating positive change. Drawing a straight line between a documentary film and a progressive social change is frequently claimed, but seldom proven as we have seen many scholars (Winston, Aufderheide, Nichols) point out. As a result, greater efforts are being made to measure and even predict the impact of documentary films by government and industry agencies, by funders of documentary projects, and even by the filmmakers themselves, before, during, and after film production. The relatively new sub-genre of ecocinema provides techniques that could be adopted by non-environmentally themed documentary films to augment its social change intentions.

While the affordances of the digital domain have yielded some impressive new theories of documentary practice and audience reach and engagement, it has also given rise to new ways of measuring the impact of documentary films and enhancing their ability to do so.

Bibliography

Ahn, Claire. *Visual Rhetoric in Environmental Documentaries.* Vancouver: University of British Columbia, 2018. Print.

Cairns, Graham. *The Architecture of the Screen: Essays in Cinematographic Space.* Bristol: Intellect, Ltd., 2013. Print.

Cavell, Stanley. *The World Viewed: Reflections on the Ontology of Film* (Enlarged Edition). Cambridge, MA: Harvard University Press, 1979. Print.

Chen, Mei-Fang. "Impact of Fear Appeals on Pro-environmental Behavior and Crucial Determinants." *International Journal of Advertising* 35, no. 1 (2015). Web. Accessed July 31, 2019. https://doi.org/10.1080/02650487.2015.1101908.

Deleuze, Gilles. *Cinema 1: The Movement-Image*. Minneapolis: The University of Minnesota Press, 1986a. Print.

———. *Cinema 2: The Time-Image*. Minneapolis: The University of Minnesota Press, 1986b. Print.

Guattari, Felix. *Chaosmosis: An Ethico-aesthetic Paradigm*. Translated by Paul Bains and Julian Pefanis. Bloomington and Indianapolis: Indiana University Press, 1995. Print.

———. *The Three Ecologies*. Translated by Ian Pindar and Paul Sutton. London and New York: The Athlone Press, 2000. Print.

Heidegger, Martin. "The Origin of the Work of Art." In *Off the Beaten Track*, ed. and trans. Julian Young and Kenneth Haynes. Cambridge: Cambridge University Press, 2002. Print.

Ivakhiv, Adrian J. "An Ecophilosophy of the Moving Image: Cinema as Anthrobiogeormorphic Machine." In *Ecocinema Theory and Practice*, ed. Stephen Rust et al. New York and London: Routledge, 2013a. Print.

———. *Ecologies of the Moving Image: Cinema, Affect, Nature*. Waterloo: Wilfrid Laurier University Press, 2013b. Print.

MacDonald, Scott. "Toward an Eco-Cinema." *Interdisciplinary Studies in Literature and Environment* 11, no. 2 (Summer, 2004). Print.

———. "The Ecocinema Experience." In *Ecocinema Theory and Practice*, ed. Stephen Rust, Salma Monani, and Sean Cubitt. London: Routledge, 2013. Print.

Moore, Jason W. *Anthropocene or Capitalocene? Nature, History, and the Crisis of Capitalism*. Oakland, CA: PM Press, 2016. Print.

O'Flynn, Siobhan. "Designed Experiences in Interactive Documentaries." In *Contemporary Documentary*, ed. Daniel Marcus and Selmin Kara. New York: Routledge, 2016. Print.

Renov, Michael, ed. "Towards a Poetics of Documentary." In *Theorizing Documentary*. Los Angeles: The American Film Institute, 1993. Print.

Rust, Stephen, Salma Monani, and Sean Cubitt, eds. *Ecocinema Theory and Practice*. New York and London: Routledge, 2013. Print.

Salazar, Juan Francisco. "Anticipatory Modes of Futuring Planetary Change in Documentary Film." In *A Companion to Contemporary Documentary Film*, ed. Alexandra Juhasz and Alisa Lebow. Oxford: Wiley, 2015. Print.

Schwarze, Steve. "Environmental Melodrama." *Quarterly Journal of Speech* 92, no. 3 (August, 2006). Print.

Seymour, Nicole. "Irony and Contemporary Ecocinema: Theorizing a New Affective Paradigm." In *Moving Environments: Affect, Emotion, Ecology, and Film*, ed. Alexa Weik von Mossner. Waterloo: Wilfrid Laurier University Press, 2014. Print.

Shembel, Daria. "DJ Spooky and Dziga Vertov: Experimental Cinema Meets Digital Art in Exploring the Polar Regions." In *Films on Ice*, ed. Scott

MacKenzie and Anna Westerstahl Stenport. Edinburgh: Edinburgh University Press, 2016. Print.
Szerszynski, Bronislaw. "The Post-ecologist Condition: Irony as Symptom and Cure." *Environmental Politics* 16, no. 2: The Politics of Unsustainability: Eco-Politics in the Post-Ecologist Era, 2007. Print.
Weik von Mossner, Alexa. "Ecocritical Film Studies and the Effects of Affect, Emotion, and Cognition." In *Moving Environments: Affect, Emotion, Ecology, and Film*. Waterloo: Wilfrid Laurier University Press, 2014. Print.
Willoquet-Maricondi, Paula. *Framing the World: Explorations in Ecocriticism and Film*. Charlottesville and London: University of Virginia Press, 2010. Print.

Filmography

A Sixth Part of the World, Director, Dziga Vertov. Moscow: Sovkino, 1926.
Berlin: Symphony of a Metropolis, Director Walter Ruttman. Berlin: Deutsche Vereins-Film, Les Productions Fox Europa, 1927.
Chasing Coral, Director, Jeff Orlowski. Exposure Labs, 2017.
Chasing Ice, Director, Jeff Orlowski. Exposure and Diamond Docs, 2012.
Gasland, Director Josh Fox, International WOW Company, 2010.
Life After People, Director, David de Vries. Flight 33 Productions, 2008.
March of the Penguins, Director, Luc Jacquet. National Geographic Films (as National Geographic Feature Films), Bonne Pioche, Wild Bunch (in Association with), Alliance de Production Cinematographique (APC) (Co-production), Buena Vista International Film Production France (with the Participation of), Canal+ (with the Participation of), L'Institut Polare Français Paul-Émile Victor (in Association with), 2005.
Nanook of the North, Director, Robert Flaherty. Les Frères Revillon, 1922.
Racing Extinction, Director, Louie Psihoyos. Oceanic Preservation Society, 2015.
Romance of the Far Fur Country, Director, H.M. Wyckoff. London: The Hudson's Bay Company, 1920.
South, Director, Frank Hurley (Uncredited). Australia: Ralph Minden Film, 1919.
Terra Nova: Sinfonia Antarctica, Director, Bob McGrath. Conceived and Composed by DJ Spooky/Paul D. Miller. No Production Company, 2008.
The Antarctica Challenge: A Global Warning, Director, Mark Terry. Toronto: Polar Cap Productions, Inc., 2009.
The Island President, Director, Jon Shenk. ITVS, 2012.
The Plough that Broke the Plains, Director, Pare Lorentz. Resettlement Administration, 1936.
The Polar Explorer, Director, Mark Terry. Toronto: Polar Cap Productions II, Inc., 2010.

CHAPTER 5

The Documentary's Digital Turn

5.1 Introduction

We have seen how the documentary film has evolved from its humble beginnings as an educational tool to an instrument of the state to promote such agendas as immigration stimulation and social reform. We have also examined how changes in form and technology introduced new theories of production and exhibition that fuelled the documentary's facility for influencing social change. As Bill Nichols observed in his book *Blurred Boundaries: Questions of Meaning in Contemporary Culture*, documentary films "can and should alter the world or our place within it … they can effect action and entail consequences" (Nichols, 67). Today, advancements in the digital domain are emerging, providing further enhancements and affordances for the documentary film to reach many more people and result in ways of exerting even greater influence on social changemakers.

A revolution of change has been introduced to the post-modern documentary. Almost every aspect of documentary film production is different in the digital domain: the technology (previously cumbersome cameras have been replaced by tablets, phone cameras, and webcams), distribution (social media, file sharing, online video-on-demand, and streaming networks), form and style (multilinear, interactive, collaborative), exhibition (laptops and mobile devices instead of cinemas and televisions), and audience (instant engagement with simulcast chat, sharing, posting comments,

M. Terry, *The Geo-Doc*, Palgrave Studies in Media and Environmental Communication,
https://doi.org/10.1007/978-3-030-32508-4_5

and star ratings; content contribution; specific, targeted audiences of changemakers). Even the language of documentary filmmaking is being overhauled: documentary films are now i-docs and web-docs; audiences are being defined as users more than viewers or spectators; screens are now monitors and platforms; cinemas are websites; distribution has become dissemination; interview subjects are now participants and collaborators; exhibition is now posting and streaming; video editing is now online editing and segment selection; out-takes are fragments; vaults are clouds; and ultimately, the filmmaker has become a digital storyteller.

The word "revolution" is appropriate in this period of the documentary's evolution. Never before in its 120-year history has so much in the documentary film industry changed so concurrently; never before have so many players in the art and commerce of documentary film production needed to adapt and continue to evolve in order just to keep up; never before has the privileged world of documentary production skill and talent been so democratized. The digital world is not simply a different world with a unique set of rules, practices, and language we can simply learn and eventually acclimatize to, but rather one of constant change, one in which both filmmaker and audience need to evolve continually, and one in which the structural mediation of the traditional documentary continues to occur.

While some (much) adjustment needs to be made, the digital documentary has opened doors and paved new roads to social change in unprecedented fashion. Just as we have discovered that participatory documentaries presented directly to audiences of changemakers can be a proven technique for successfully achieving change, we are now finding that new advances afforded by digital ecologies are enhancing the experience for both filmmaker and changemaker, ameliorating the path to social change. This chapter will examine some of the ways this (r)evolution of the documentary film are taking place today and which methods and affordances work best in improving the documentary's ability to impact on social change.

5.2 Digital Documentary and Social Change

Historically, the documentary film has experienced much use and success in informing the public and influencing public opinion. In their essay, "Documentary Film: Towards a Research Agenda on Forms, Functions, and Impacts," Michael C. Nisbet and Patricia Aufderheide identify the documentary's newly evolved purpose of social activism in the digital domain:

> Documentaries are no longer conventionally perceived as a passive experience intended solely for informal learning or entertainment. Instead, with ever increasing frequency, these films are considered part of a larger effort to spark debate, mould public opinion, shape policy, and build activist networks. (Nisbet, Aufderheide, 450)

Social issue documentaries released online have the capacity and the ability to influence the masses and mobilize them to effect social change, and, indeed, there are some recent examples of success in this area—such as the movie *Blackfish* that resulted in plunging SeaWorld, a marine theme park, into crisis mode, forcing them to reconceptualize their exhibits and entire corporate philosophy—but what happens when you remove the masses and the need for emotional engagement? How do policymakers interact directly with the digital documentary film?

Scholars (Waugh, Nichols, Marchessault) agree that the documentary film has developed into a relatively influential tool for social change and one that is growing in power, with technical advancements being made on a regular basis in the digital domain. Ezra Winton and Svetla Turnin, co-founders of Cinema Politica, a non-profit media arts organization specializing in showcasing activist documentaries, have found a reciprocal arrangement flourishing between the documentary and its reformative ambitions: "Just as a social movement could inspire the making of a documentary, a documentary in turn could activate a social movement" (Winton, Turnin, 21). This illustrates how activation is now being provoked, thanks to, most evidently, social media campaigns involving documentary films. The contemporary documentary often involves activists in the production's fund-raising process before the film is made and continues with momentum long after the film is released. The philosophy of Nichols' participatory mode of documentary filmmaking has now extended to audience engagement before and after the documentary viewing experience, and with such digital measurement programmes as Harvis, sometimes even during the documentary viewing. With a greater potential audience reach in the digital domain, there is the possibility of many more people rallying behind a documentary film's social issue by participating in these activist movements to contribute directly to the film's social change agenda.

This recognized ability to influence policy and cause social change, especially in the digital domain, is now involving traditional filmmaking industry partners. Conventional buyers of documentary films—television networks—have recognized the importance and value of digital distribution not just in terms of making direct sales of the titles they present, but also as a promotional tool to build audiences by supporting online activist campaigns related to the films. In America, the Public Broadcasting System (PBS), for example, provides a comprehensive list of "Engagement Strategists"[1] on its website, encouraging independent filmmakers to work with these activist organizations not only to achieve their project's social issue goals, but also to raise public awareness of their "product."

But while these strategies and tactics may be effective for the traditional, linear documentary film in its online existence, does this apply as well to the contemporary, multilinear, interactive, database, and digital documentaries? In his book, *The Language of New Media*, digital media theorist Lev Manovich indicates an inherent flaw intrinsically attached to contemporary documentary production and related specifically to the constantly evolving new media technologies and the digital domain in which they exist:

> Creating a stable new language (in digital film) is … subverted by the constant introduction of new techniques over time … Every year, every month, new effects find their way into new media works, displacing previously prominent ones and destabilizing any stable expectations that viewers have begun to form. (Manovich, 243)

While the art form continues to evolve, the audience struggles in trying to keep up with an ever-changing set of parameters of engagement. It is simultaneously constructive and destructive. In the second edition of *New Challenges for Documentary*, Alan Rosenthal and John Corner address this struggle in their introduction:

> The non-linear, interactive properties of the web create possibilities for documentary work at the same time as they are an obstacle to the kind of public 'confrontation' of viewer with topic that many documentarists have sought. In this confrontation, it is often quite important that viewers do not have too many options immediately available for negotiating their route through

[1] "Engagement Strategists," *pbs.org*. Accessed September 2, 2015. http://www.pbs.org/pov/filmmakers/engagement-strategists.php#.VedcF_lViko.

> the material, that they just have to 'look and listen', so to speak. (Rosenthal, Corner, 5)

The concept of "too many options" is at the heart of the digital documentary. Supporting media, often referred to as "metadata," such as additional video, hypertext, web-links, and pictures, may provide an enhanced presentation of data related to the documentary's subject matter, but engaging in them interrupts the traditional experience of simply "looking at and listening to" a linear documentary. While the concept of too many options seems to be a hindrance to the documentary viewing experience in the digital domain, some scholars believe these options are carefully created additions that actually enhance the viewing experience. Florian Thalhoffer, one of the creators of *Korsakow*, a computer programme that makes video fragments of a documentary story available for the viewer to interact with (to be examined in detail in Chap. 6), said in an interview with Kate Nash that "(t)he viewer has this option or that option, but these options are really pathways that have been pre-thought and planned by the author" (Nash, "An Interview," 193). He goes on to acknowledge that while this method of documentary storytelling provides more information in a different way, there is still resistance to it as we struggle to learn and become comfortable with this new approach.

> The beauty of computers is that they can free you from this linear way of thinking (traditional documentary films). But it is difficult; we don't really understand how to do it. We have a lot of knowledge about how to tell linear stories, more than 120 years of film history in fact. We're still learning how to tell stories in a non-linear way. (Nash, "An Interview," 193–194)

Nevertheless, the digital documentary has benefitted in areas such as extended audience reach and engagement, despite these technical obstacles of user interface. By making, exhibiting and disseminating documentaries in this new medium, the genre's ability to influence audiences is magnified, thereby increasing its potential for causing social change. Audience engagement, in particular, is one of the most significant affordances provided by the digital domain for the documentary film, so much so that a new kind of documentary film has emerged: the i-doc.

5.3 The Interactive Documentary (or i-doc)

One of the first changes in the remediation of any medium is often the name itself. Several new identifiers have emerged to describe more accurately and categorize the digital documentary. Canadian documentary filmmaker and winner of the 2006 Governor General's Award in Visual and Media Arts, Peter Wintonick (*Manufacturing Consent: Noam Chomsky and the Media*, 1992), abandoned the word "documentary" altogether referring to the more collective term "docmedia" which encompasses many digital documentary forms and their respective terms: web-docs, transmedia docs, database docs, and the one this section of this chapter will examine, the interactive documentary or i-doc (Winston, xv).

The interactive nature afforded by the digital domain gives the i-doc a relational significance between the filmmaker and audience that is more profound than previously experienced between the two when the film's exhibition was restricted to a cinema or television screen. The i-doc can be viewed anywhere at any time with a laptop or mobile device. It can also provide instant feedback from the audience to the filmmaker through interconnected platforms such as comments pages and chat rooms. These unique affordances of time and place extend beyond the traditional encounter between a documentary's director and their viewers. New relationships between the two are developing at the pre-production, production, post-production, and release stages of an i-doc project in ways the analogue documentary filmmaker could only have dreamed of.

Interview subjects and targeted audience groups are now stepping up with funding in the pre-production phase, taking on the role of the documentary's executive producer. This raises some obvious ethical issues wherein the profiled group or the changemaker may be granted the authority to control content and messaging resulting in propaganda and misinformation; however, while the possibility for the tail to wag the dog exists, many such financial arrangements entrust the filmmaker with the storytelling expertise within the medium. For example, documentary director Harold Crooks (*The Corporation*, 2003) was commissioned by Quebec policy adviser Brigitte Alepin, author of *La Crise Fiscale qui Vient* (*The Coming Fiscal Crisis*), when she wanted to see a documentary film made on the subject of her book. In addition to being a Quebec tax policy adviser, Alepin is also an accountant and anti-tax-avoidance crusader. Acknowledging her lack of experience in film production, she hired a director who was not only competent, but also an expert in documentary research in corporate crime. Alepin appears in the film *The Price We Pay*, which she co-wrote with Crooks. She also interviewed many of the film's francophone subjects (Gray, *The Globe and Mail*, 2017). We now see an evolution in the participatory process that includes the policy adviser, not just as interview subject and targeted audience member, but as filmmaker and activist.

The policy adviser's goal is to create financial policy in order to institute progressive economic change. *The Price We Pay* was screened exclusively for heads of state worldwide, wherever possible, ahead of film festivals, the traditional first venue of exhibition for a film. In France, a private screening was arranged for Finance Minister Michel Sapin who "lauded the film as *excellent* on national television. … and called the tax practices it highlights *absolutely scandalous*" (Gray, *The Globe and Mail*, 2017).

In this case, we are seeing the government collaborator as activist, using the filmmaker to create a documentary that delivers a message of desired social change not only in their government, but in governments throughout the world. The unusual distribution strategy bypasses the general public and aims its sights directly at international tax policy changemakers. But not all pre-production interactivity involves funding. While Alepin hired Crooks to make the documentary on a subject in which she wanted to see policy change result, other changemaking audiences request the production of documentary projects and suggest particular content to be addressed in order to suit their needs for informed policy creation.

The films of George Ferreira address the need of Industry Canada to identify socio-economic issues of the aboriginal Keewaytinook-Okimakanak communities in Northwestern Ontario, an area too remote for policymakers to visit regularly. Instead of the previous documentary process of presenting film content solely from the bottom-up, we now have an approach that incorporates top-down content requests from the changemaker to the filmmaker. The additional affordance of Internet access to the content makes the address to the issues more immediate, allowing for desperate funds to be allocated in a more expedient manner. Citing the Fogo Process as an inspiration, Ferreira argues that the i-doc advances the goals of social reform:

> Streaming video, in particular, allowed participants to revise and contribute to the editing process remotely. It also meant that videos could be seen by senior bureaucrats within days, and frequently within hours, of being produced … in particular, video applications combined with broadband, helped

> create the context whereby the Fogo Process could be applied in a new and innovative way. (Ferreira et al., 92)

5.4 The Collaborative Documentary

The participatory documentary is redefining itself more as a comprehensive collaboration between various stakeholders and audiences in the digital domain as more engagement tools develop in the online production, exhibition, and dissemination of these films. Documentaries are no longer simply stand-alone one-offs viewed by a passive audience, casually discussed and then forgotten. More and more, the audience is collaborating with the filmmaker as an active participant not only to request and suggest content, but to provide it as well. Many of these documentary projects are now considered "living," a film without end viewed and, at times, made by a global audience. In her essay, "Strategies of Participation: The Who, What and When of Collaborative Documentaries," Sandra Gaudenzi describes this as a new form of documentary:

> The ability to upload content to an online documentary gives it a fluid form that is not achievable in a linear documentary. Since new content can potentially be uploaded *ad infinitum*, the collaborative documentary becomes a constantly mutating entity. What could now be seen as a *living documentary*. (Gaudenzi, "Strategies of Participation," 130)

Film audiences have never been more empowered to control their own content, and while this bodes well for consumers of YouTube videos of pet antics and wardrobe malfunctions, how effective is this process in influencing the changemaker? We have seen that when the people of Fogo Island or the Keewaytinook-Okimakanak communities of northern Ontario or even the world's climate science community collectively participate in a documentary project that is targeted directly at an audience of changemakers, the result is often positive: their aims met, their goals achieved. What they all seem to have in common is a factual and first-person presentation of the story from a source the audience considers reliable, authoritative, and trustworthy, something not all collaborative documentaries masquerading as the visualization version of a petition can claim. But does this approach ensure success?

While audiences/users are being engaged more than ever before, according to Kate Nash in her essay "Clicking on the World: Documentary

Representation and Interactivity," "there is no necessary connection between interactivity and audience empowerment" (Nash, "Clicking on the World," 50). She sets a framework for examining the interactive documentary conceived as multidimensional "in which the actions of users, documentary makers, subjects and technical systems together constitute a dynamic ecosystem" (Nash, *Clicking on the World*, 51).

Documentary theorist Jon Dovey echoes this metaphor in his own analysis of the importance of collaboration in producing interactive documentaries: "Thinking ecologically suggests we look at big pictures, at the whole assemblage of agents that constitute documentary ecosystems" (Dovey, 11). These digital agents work together in a symbiotic manner to maintain a new kind of participatory culture in documentary production and engagement characterized more commonly as collaboration. Dovey identifies four key agents in this ecosystem:

> '**Affiliation**', elective group formation in online community around enthusiasms, issues or common cultures; '**Expression**', music, video, and design tools in the hands of far more users than ever before and being used for every kind of human mode of communication; '**Collaborative Problem-solving**', mobilizing collective intelligence, crowdfunding, online petition making, alternate reality gaming, wiki-based shared knowledge practices; and '**Circulations**' playing an active role in directing media dynamics through the new flows of viral media driven by Twitter, Facebook and YouTube. (Dovey, 12) (*Emphasis is mine*)

These four affordances of the interactive documentary ecosystem remediate the film genre in ways that give new agency to the filmmaker and their stakeholders. Within this new digital ecology thrives a new citizenry comprised of filmmakers, researchers, interview subjects, spectators, and changemakers, all collaborating in this unique participatory culture.

- **Affiliation**: Previously identified predominantly as the interview subject in a documentary film, these participants have frequently been members of an under-empowered group requiring the voice provided by the filmmaker. The former participatory mode enhanced and clarified their representation. Through the more collaborative agency of affiliation, they involve more players in ways that are concurrent to production, sometimes even initiating projects.

- **Expression**: The filmmaker is no longer the sole creative force behind the interactive documentary. In collaboration with the production's participants and stakeholders, including audiences, elements such as music and metadata can be contributed to the project providing a more focused representation of the profiled subjects, enhancing their trustworthiness as their contributions are coming from them directly, not the filmmaker indirectly. In this collaborative ecosystem, the filmmaker is not replaced, but given a new role of using their own expertise to incorporate these artefacts of expression within the documentary project.
- **Collaborative Problem-solving**: Certain issues require certain approaches to best serve the changemaker. It is often more effective to convey video reports directly from the subjects of a documentary than it is to present video reports from angry protestors demanding the same change, but who are speaking on behalf of the needy party. Sometimes the opposite is more effective. Knowing your audience is essential in selecting the best digital tools and practices to serve the specific needs of the changemaker. Content and data related to the issue are equally important in influencing the changemaker so that they may be fully informed. Collaborative research from all interactive documentary stakeholders assists in this project. Another "problem" that collaboration in this area helps solve is production financing. Both at the beginning of a project and at its end, production funds to produce and to promote are being raised through online campaigns such as GoFundMe and Indiegogo, the primary supporters being the stakeholders within the project's ecosystem.
- **Circulations**: More on the traditional distribution and exhibition side of the documentary, sharing and posting on social media sites in particular provide a new expansive approach to having a documentary film seen. In addition to this viral method of screening, collaborative feedback from viewers further augment the film's message through additional content as well as the film's goals with contributions from changemakers and related stakeholders.

Theoretically and ideologically, this collaborative process is an attractive new methodology for documentary filmmakers, yet there still exists some resistance to its practice among professional documentarians and activists alike. In her chapter, "User Experience Versus Author Experience," Sandra Gaudenzi suggests three reasons for this resistance:

> One could make a case that it is just a question of time, and that changing workflows in teams and institutions is a slow process. One could also point out that interviewing users and testing prototypes does involve a cost, and that small productions cannot afford the process. To these two valid explanations I would like to add a third one: that for many documentary makers there is resistance to the idea of testing their work. (Gaudenzi, "User Experience," 124)

Invoking my professional experience as a filmmaker in both fiction and non-fiction formats, I would also like to add to this. I believe there exists a resistance to collaborate from the filmmaker for artistic considerations. I have seen entire productions, fully financed and with locked production schedules, collapse due to the proverbial "artistic differences" that surface from the inexperienced partner who now stakes a claim in the creative approach and content. The privileged artistic vision of the director has now been compromised by competing and often conflicting visions from non-filmmakers.

This notion has been addressed by former head of the National Film Board of Canada Tom Perlmutter in a 2014 article he wrote, entitled "The Interactive Documentary: A Transformative Art Form." To the documentarian who will not be "dictated" to by their audience, Perlmutter says:

> To take this attitude is to misunderstand profoundly what understanding audience means in an interactive world, where as creator you make the audience a collaborator in your processes. This does not invalidate the filmmaker as creator or auteur. It enlarges the notion of auteur. The new auteurs will understand that the relationship to audience as co-creators and collaborators is part of their medium of creation. (Perlmutter, "The Interactive Documentary")

Adapting to the new forms and practices of the interactive documentary and its collaborative community of creators and users is at once both eagerly embraced and reluctantly accepted. The reason both sides engage with it is because the common goal of informing and influencing social change has never been greater. The addition of new tools, techniques, and stakeholders now provides the changemaker with more information in more immediate fashion, enabling them to enact progressive change faster and with more accuracy. And the activist filmmaker now has an army of collaborators at the ready to contribute to their messaging through content and dissemination. Even the elusive measurement of impact a documentary has on the social issue it presents is being made easier and more precise with digital analytics and related affordances.

5.5 Impact Measurement

As examined in previous chapters, many scholars in film studies, political science, and communications have theorized that the documentary film indeed has the capacity to cause change by illuminating its subject and providing visual context to the profiled issue, and the definitions these theories provide help us seek direction as to how to use this persuasive communications tool to achieve this goal. Perhaps the most accurate of these definitions come from American political scientist David Whiteman:

> Producers and activists seeking to maximize political impact, and scholars seeking to understand political impact, benefit from conceptualizing the production and distribution of a social issue documentary as an intervention into a policy process. (Whiteman, "The Evolving Impact of Documentary Film")

The key word here is intervention. It is a very appropriate word for the purpose of defining the relationship between film and policy because it often means action taken to improve a situation or "to intentionally become involved in a difficult situation in order to improve it" (*Cambridge Dictionary*).

American documentary filmmaker George Stoney, who worked with the NFB and the *Challenge for Change* film series programme and, in particular, as an executive producer on the *VTR St-Jacques* film examined earlier, adds to this conviction with an even more specific definition focused on the duties and obligations of the documentary filmmaker:

> 50 percent of the documentary filmmaker's job is making the movie, and 50 percent is figuring out what its impact can be and how it can move audiences to action. (Karlin, Johnson, "Measuring Impact")

This puts an onus squarely on the shoulders of the filmmaker to serve the goals of the film by intervening as an activist after their craft labour is complete. But how successful is the filmmaker in this regard? There are many sceptics of the impact of the documentary film's ability to influence social change.

In a 2014 report by Patricia Finneran entitled *Documentary Impact: Social Change Through Storytelling*, released by Hot Docs, the popular Toronto-based documentary film festival, this claim is supported: "(Impact) can often be nuanced and difficult to quantify. Even with hindsight, understanding of what created change is always contested" (Finneran, 5).

But is social change the only measure of impact for a documentary? According to another report that same year, authored by Caty Borum Chattoo and Angelica Das for the Center for Media and Social Impact in Washington, DC, the impact of a documentary film is defined as "change that happens to individuals, groups, organizations, systems, and social or physical conditions. Typically long-term and affected by many variables, impact represents the ultimate purpose of community-focused media efforts—it's how the world is different as a result of (documentary film) work" (Chatoo, Das, 1).

In the few cases where the connection between the documentary film and a social change can be clearly identified—the Fogo films, the films of the Keewaytinook-Okimakanak communities examined in Chap. 3, and to be discussed later, *Sin by Silence* and *The Polar Explorer*—the audience was not a group, an organization, or the general public. The specific audience that yields the direct connection between a documentary film and social change is the changemaker.

The Washington report suggests that "many variables" contribute to the documentary film's success at achieving impact when it does, once again muddying the waters in identifying to what degree a documentary film can claim to be the cause of social change. However, despite—or perhaps because of—this scholarly scepticism, new digital tools and funding agencies are emerging to assist activist filmmakers in measuring the impact of their films before, during, and following production. Philanthropic funders such as The Fledgling Fund, BRITDOC, and the Documentary Australia Foundation have all introduced programmes that assist the filmmakers they financially support in activism campaigns and analysing the results. All of these programmes rely on the affordances of the digital domain, affordances that accelerate and invigorate the social change aspirations of the documentary film.

Impact measurement, as it is currently being defined by digital documentary practitioners and stakeholders, serves two purposes: identifying success politically in achieving change and identifying achieving goals as defined by documentary funders. Meg McLagen explains the difference between the two in her chapter "Imagining Impact: Documentary Film and the Production of Political Effects":

> On the one hand, (the term impact measurement) refers to demonstrable political effects something can have in the real world, such as, for instance, helping to raise money for new schools in the developing world, as the film *A Small Act* did in Kenya in 2010. On the other hand, the term refers to the institutionalization of audit practices through the introduction of a set of concrete performance criteria by which such change can be imagined and then assessed. In other words, for social entrepreneurs, much like money managers, the key issue is to invest in socially valuable projects that can provide quantifiable returns. (McLagan, 309)

As difficult to quantify as it is, there is a movement among the industry, its artists, its funders and even government partners to create measurement tools designed to be used before and after a film's production to ameliorate and enumerate the film's success as a social change agent. As McLagan points out, these measurements are often used to assess funding for these projects. In fact, many new and emerging funding agencies are expecting filmmakers to project the impact of their documentaries as a condition of financial support. As a result, the new digital tools and tactics that are being created for both filmmaker and funder are, at this stage, largely experimental, but are demonstrating varying levels of success in establishing and identifying that elusive connection between a documentary film and social change results.

Along with these new technologies and practices comes a new position for the tail credits of documentary film: impact producer. The Documentary Organization of Canada defines *impact producing* as a new term

> to describe a new space in which filmmakers are mobilizing people, networks, and resources to create change. Impact producing is a very comprehensive way of grouping together and honing the many tasks that enter into the process of the making and marketing of a successful documentary … it is a burgeoning field of skill development that upon closer scrutiny is incredibly rich in all of its possible permutations. (*Impact Report and Toolkit*)

Occupying a large area of this new space is the Fledgling Fund, operating in New York City since 2005. While its focus on impact relates to audience engagement, they do support some initiatives that are aimed directly at changemakers. Rather than simply being a grantor, the Fund acts as a partner supporting production, strategic outreach, and audience engagement campaigns. The Fund provides its own resources (databases) as well as specific strategies that have worked in the past for social issue documentary films with similar themes. According to a 2015 report by Ann McQueen, nearly \$12 million has been distributed to more than 330 documentary film projects from an endowment of around \$15 million (McQueen, *The Fledgling Fund*).

Access to Big Data was not so easily available to the documentary filmmaker prior to the Internet, so now with online databases like those provided by the Fledgling Fund, the activist documentarian can extend their film's reach not only to a global audience, but also to targeted audiences of changemakers, facilitating their goal of achieving social change.

The organization places a strong emphasis on collaboration in order to establish the best methods of production and post-production audience engagement strategically customized to maximize potential success, as detailed in this report by Emily Verellen:

> (O)utreach and strategic communication is largely determined by how the film fits into the social movement, how the movement itself has connected with the film, embraced it and worked with the filmmaker to understand the message it conveys, how it fits into the needs of the social movement and how the members of this movement can see it. In order to do this effectively, film teams (made up of filmmakers, outreach and engagement coordinators, movement builders and/or leaders/organizers) have to think critically about how and where the film's message should be conveyed. (Verellen, 6)

In order to facilitate this process, the Fund makes available to its participants online programmes and digital tools that include access to subject-specific research papers and case studies so the filmmaker can compare the success and failure of other campaigns and strategies and apply these data and methods to their own initiatives. Access to the Fund's own databases which provide information on changemakers is also available. Access to this archival information online accelerates the production process as well as its activist intentions: digital data supporting digital documentary and digital activism.

In addition to filmmakers, the Fund financially supports organizations that focus on impact measurement. One such project is the Tribeca Film Institute's *Interactive Media Impact Working Group*. Created with the help of MIT's Open Documentary Lab, the group comprises of "an interdisciplinary group of artists, funders, social scientists, academics, impact producers and activists to explore how interactive non-fiction projects can and should be measured in terms of impact and engagement," according to Ingrid Kopp (Kopp, *Interactive Media Impact Working Group*). This kind of collaboration or joint-venture approach to activist documentary filmmaking was not possible before the world was networked through the World Wide Web. Partnerships were made, but not on this scale and not in a relatively brief period of production time.

Digital documentary theorist Nicole Marie Nime addresses this advancement in her paper "The Impact of Digital Technology on Documentary Distribution":

> Documentary has fallen into a period of crisis. It also has embarked upon an era of growth. The digital age has engendered this paradox. The Internet has opened new avenues of exhibition, causing mainstream media institutions, which traditionally have monopolised access to audiences, to progressively lose control of their most valuable assets. It also has created the opportunity for filmmakers to instantly distribute their work to global audiences and develop direct relationships with these individuals. (Nime, 11)

These direct relationships with targeted audiences assist the social issue documentary film in reaching the changemaker and increase its chance of achieving the social change it strives for, while, at the same time, undermining its primary revenue stream, television, referenced by Nime as "mainstream media institutions." With digital distribution, a previously untapped network of online exhibition is now open to the filmmaker, thereby devaluing the licence fees they used to receive from the exclusive broadcast arrangements they had with television networks. The social change goal is heightened at the cost of its primary revenue source.

How exactly does this increased access to audiences result in influencing social change objectives for the documentary film? The extended reach of not only more viewers is obvious, but also the digital technology allows for reaching wider bases of targeted audiences, those empowered to enact change. An example of how this process now works in the digital age is the 2009 film *Sin by Silence*.

5.6 Case Study: *Sin by Silence* (2009)

This forty-nine-minute documentary film by Olivia Klaus about domestic abuse benefitted from support from the Fledgling Fund and campaigned successfully to introduce new legislation on domestic abuse. The film title is even mentioned in the name of the bills—*Sin by Silence Bills*—that were passed into California law in September 2012 (Case Study: *Sin by Silence*), establishing that elusive connection between a social issue documentary and the social change it set out to achieve.

The film focuses on an advocacy group called Convicted Women Against Abuse (CWAA), the first inmate-initiated and inmate-led group in the history of the US prison system (Klaus 2009). Members of the group tell their stories, providing a unique perspective on the issue of domestic violence from inside the California Institution for Women.

These are the objectives of the film's impact campaign:

- Create awareness and conversations about the silent tragedy of domestic violence.
- Build connections and mobilize individuals around the domestic abuse prevention movement, with a focus on local resources for involvement.
- Educate communities about the effect of domestic violence and the urgent need for prevention and change.
- Help influence others to join the movement. (Case Study: *Sin by Silence*, http://www.thefledglingfund.org/)

One of the women profiled by the CWAA in the film is Norma Cumpian, a forty-year-old victim of domestic violence who was pregnant when she killed her boyfriend during an attack. The courts did not buy her self-defence claim and sentenced her to fifteen years. It was during this time that she joined the CWAA. The group campaigned on her behalf introducing new evidence not presented at her trial. The filmmakers featured her story in *Sin by Silence* and launched an elaborate national campaign for her freedom as well as the freedom of other women with similar stories.

The filmmakers began their campaign by identifying the US states with the worst domestic violence statistics: Arizona, California, Indiana, Kansas, Louisiana, Missouri, New Mexico, Ohio, Texas, and Washington. The *Stop the Violence* tour travelled throughout these ten states, had forty screenings, and connected with more than 5400 screening attendees. In each state, organizers partnered with local organizations and non-profits with a focus on domestic violence to expand their audience reach.

The filmmakers also made two hours of teaching videos to spark further discussion. Digital versions of these resources were made available to educators and advocates and available on their website. Each film contained a call to action related to the proposed legislation. One such video asset specifically calls on Governor Jerry Brown to pass the bills (*Urge Gov Brown to Pass AB 593 & AB 1593*, 2012, youtube.com).

The Fledgling Fund report lists many impacts directly related to the film and its outreach campaign. Here is a sample of one of them:

> Students started a campaign to help free Norma Cumpian (the featured subject in the documentary) after spending 18 years behind bars for defending the life of her unborn son against her abusive boyfriend. The campaign collected over 2000 signatures and viral videos circulated online. Norma was released from prison in November, 2011. The (California) Attorney General acknowledges the impact of public support (from the film) for her release as the main reason that influenced the Governor's decision. (Case Study: *Sin by Silence*)

Documentary scholars such as Heather McIntosh acknowledge that the film is directly responsible for correcting this social injustice: "What started as a documentary film telling these women's stories and raising awareness about domestic violence and its myths grew into a movement that resulted in changes in California state legislation" (McIntosh, 223). It is important to note here that without the Attorney General's acknowledgement and the name of legislation bearing the title of the film—the *Sin By Silence Bills*—there would be no direct connection, or proof as Winston and Finneran might say, of the film's influence and impact on the release of Ms. Cumpian from prison.

5.7 BRITDOC (UK): Founded in 2005

This chapter has discussed the digital tools now available to filmmakers that documentarians before the Internet did not have access to. Many of these online tools are aimed at assisting the activist filmmaker in measuring the impact of their projects. One of the largest impact measurement organizations in the world is BRITDOC whose motto is "Dedicated to the

Impact of Art and the Art of Impact" (*About BRITDOC*, BRITDOC.org). Representative of that dedication is the *Impact Guide.*

Available for free on their website,[2] it is a step-by-step course and online resource that guides the social issue documentarian through areas of impact filmmaking with which they may not be familiar, recommending courses of action that have worked best in the past. The guide has five sections: *Introduction*, *Planning*, *Impact in Action*, *Impact Distribution*, and *Evaluating*. Within each unit is a comprehensive overview of what needs to be done to raise funds, promote messaging, and achieve activist goals.

In 2014, a division of BRITDOC, *Doc Society*, whose stated mission is "to befriend great filmmakers, support great films, broker new partnerships, build new creative models, share new knowledge and develop new audiences for documentary globally" ("About Us," *Doc Society*), launched another free online resource for filmmakers, the *Impact Field Guide and Toolkit.* This wide-ranging guide covers a lot of ground for the impact-oriented filmmaker, but one of the most interesting and valuable sections of the guide is the section *New Tools for Impact Documentary.* Here, the guide offers several digital tools for measuring impact (*New Tools for Impact Documentary*):

1. **ConText**: A contraction of "Connections" and "Texts," this tool is designed for film impact evaluation by focusing on "understanding the contribution that a film makes to the online debate on a given issue, and whether the key influencers on that issue are involved in and affected by that debate" (New Tools…). The methodology involves mapping, monitoring and analysing the social networks of stakeholders with the main topic of the documentary (the social relationships) and the content of the information shared by these agents (content) (New Tools…).
2. **Harvis**: This tool is available on mobile devices to engage audiences during a screening. A series of customizable questions can be created and shared with audience members who have subscribed to the application. A simple swiping gesture for up ("I want to take action") and down ("I feel helpless") is all the audience member needs to do while watching the film. At key points, the film stops for three minutes to allow viewers to post a comment. Audience members can also be identified as educators, stakeholders, parents, or concerned

[2] https://impactguide.org.

citizens. When analysing the collected data, these viewer types provide an additional metric of how different audiences are being impacted by the film.

3. **Media Cloud**: This open-source data platform allows film academic researchers, journalism critics, policy advocates, media scholars, and others to examine which media sources cover which stories, what language different media outlets use in conjunction with different stories and how stories spread from one media outlet to another. This approach is less audience-focused as it is media-focused providing a perspective from professional journalists on the social issue of the documentary and the documentary itself.

   The Media Cloud tool is an interesting way of mining new data related to impact. Not only does it search media online to determine language related to coverage of a social issue documentary film, it can also find references in other sectors online, such as government stakeholders who may be creating new policy based on a particular film.
4. **Ovee**: Described as a "social TV experience," Ovee is an online event involving the screening of a film and simultaneous interaction through chats, polls, emoticons, and social media with audiences, filmmakers, and other stakeholders. Hosted and maintained by Independent Television Services, the service offers private or public screening events. Throughout the live screening, Ovee gathers data regarding the composition of the participating audience and its level of engagement with the film during the screening, specifically to identify how the audience responds to the story emotionally, how it shares information, and how it takes action.

5.8 The Focus on Audience in Impact Measurement

Many of the digital tools and resources examined in this chapter to measure the impact of documentary film are aimed at audiences representing the general public. I am not convinced that the metrics related to audiences are reliable. For the most part, audience members who are being polled have willingly attended the screening, knowing in advance what the subject matter of the film will be. These audiences are largely supportive of the social issue being profiled in the first place, and their responses will reflect their bias in this regard. Naturally, they agree in large numbers to be motivated to act, but in the final analysis, where is the metric that shows that these audience members actually took action after they left the theatre? In their book *Crafting Truth: Documentary Form and Meaning*, Louise Spence and Vinicius Navarro argue that too much emphasis is placed on the general public in documentary impact measurement:

> (M)any film and video makers not only have respect for their viewers, they see them as able to process information and use it to effect transformations ... In practice things can work somewhat differently since there is no guarantee that the public will turn their newly acquired knowledge into action. Awareness does not necessarily lead to social change. (Spence, Navarro, 105)

This perspective is echoed by Daniel Marcus in his essay "Documentary and Video Activism." He argues that the limited engagement filmmakers have with their audiences makes it impossible for them to ensure their viewers will take action:

> The political and social documentary ideal ... includes an audience that responds to a viewing of the film by actively intervening in the situation depicted in the work. Yet this ideal is often a chimera. Documentary producers have a limited window of intervention; their engagement with audiences rarely goes beyond the temporal and physical limits of the viewing experience itself ... Few producers have tried to relate exhibition of their work with the activities of their viewers after the screening is finished. (Marcus, 190)

Some scholars, on the other hand, insist that audiences of the general public are the primary force in creating a groundswell that ultimately motivates the changemaker to make change. Two organizations run by documentary scholars are dedicated solely to social activist documentaries and in engaging its filmmakers, communities, and on-screen subjects: *Cinema Politica* at Concordia University in Montreal, Canada, and the *Centre for Media and Social Impact* at American University in Washington, DC.

Founded in 2001 by documentary film scholars Ezra Winton and Svetla Turnin, Cinema Politica claims to be "the largest volunteer-run, community and campus-based documentary-screening network in the world" (*About Us*, cinemapolitica.org). In its brief history, it has grown to more than one hundred screens and has a following of nearly 5000 on Facebook.

Its mandate is "to promote, disseminate, exhibit and promote the discussion of political cinema by independent artists, with an emphasis on Canadian works" (*About Us*, cinemapolitica.org), as evidenced by such works as *Honour Your Word* (2014), a film about the political injustice faced by the Algonquins of Barriere Lake, Quebec; *'Til the Cows Come Home* (2014), a film about the government's decision to transform dairy farms to prison farms[3]; and *Demur* (2011), a documentary short about the civil liberty violations during the G8 and G20 meetings in Toronto in 2010. In a recent interview with DocuDays UA, Winton explains that his organization's goal is to establish "encounters" between documentary films and audience:

> I like using the term *encounter*—that we encounter media, that we encounter art. Encounter in a sense that we are actors in a social world coming up against objects of art and forces of media and communication, and we encounter them, we came up against them. Sometimes, we are repelled by them. Other times, we are brought in closer to them because they touch us in a certain way, or we react in a way that is so meaningful to us that it changes us. We try to create social spaces that are in the material world, that are virtual for the documentary to be encountered by audiences, and that are appearing in a space that isn't structured by commercial or marketing imperatives. (*About Political Films*, docudays.org.ua)

Encounters are indeed important to the documentary film's ability to affect social change. While having the general public encounter a film and be moved to action is a laudable result, true change is measured more frequently when the audience is the changemaker. In addition to the examples previously provided, one recent example of a documentary film that mobilized its audience to cause change is Gabriela Cowperthwaite's *Blackfish* (2013). By showcasing the mistreatment of whales in an aquatic amusement park, the film has been directly attributed to causing SeaWorld

[3] According to a report made by *CBC News*, June 21, 2018, Canada's federal government budgeted $4.3 million to restore prison farms in the Joyceville and Collins Bay institutions in Kingston, Ontario. "Feds release details on plans to restore prison farm program at Joyceville and Collins Bay institutions," *CBC News*, June 21, 2018. Accessed August 24, 2018. https://www.cbc.ca/news/canada/ottawa/kingston-prison-farm-cows-goats-1.4716232. In a September 10, 2018, phone call with the office of Mark Holland, parliamentary secretary to the minister of public safety, a spokesman said there was no influence of the film *'Til the Cows Come Home* regarding the government's decision to restore prison farms.

to lose more than twenty-five million dollars (US) ("The Blackfish Effect," *The Daily Mail Online*) by its corporate executive team—one of the rare examples of a documentary's general public audience directly causing change. Even Brian Winston concedes that this is a bona fide example of social change being caused by a film's emotionally engaged audiences.[4]

In the same DocuDays interview, Ezra Winton admits that "it's really difficult" to measure the impact a documentary has on societal change, but believes his organization's open-source access to filmmakers of social activism documentaries as well as all audiences (admission is not charged) creates an opportunity for social change through Cinema Politica's encounter philosophy.

In the US, a similar organization known as the Centre for Media and Social Impact (CMSI), founded by documentary scholar Patricia Aufderheide, aims to create, study and showcase media for social impact ("About Us," http://cmsimpact.org). While presenting films for public debate in much the same manner as Cinema Politica, the CMSI focuses on developing scholarship that directly connects activist documentary projects with political action taken by its general public audiences. In a 2010 open letter to the US Federal Communication Commission, Aufderheide and Jessica Clark identified the "5 Cs" of audience engagement in today's digital landscape:

- choice,
- content creation,
- collaboration,
- curation, and
- conversation. (Clark, Aufderheide, 3)

Clark and Aufderheide argue that these actions can all be taken by today's documentary general public audience, thanks to the affordances of the digital domain. Prior to the Internet, audiences would simply attend a screening in a cinema or watch a documentary on their television, and that was the extent of their engagement. In the digital era, their participatory opportunities are quite expansive including content creation.

[4] Stated in a classroom lecture of the course *Contemporary Documentary* taught by Seth Feldman, August 18, 2015, York University, Toronto, Canada.

> Public media in the digital era escapes the traditional zones of mass-media, and therefore public media is now defined by what it does, not where it is. Public media will be public depending on the degree to which it is useful in promoting public life—engagement with the fundamental issues of the society and its choices for the shared terms of life together. (Aufderheide, Clark, 11)

They suggest that audiences today have the opportunity to shape their own society and can be directed to do so through this unprecedented engagement with digital media. Whether they do or not is still a question that requires further investigation and evidence. Aside from the *Blackfish* exception to the rule, it still seems that the greatest evidence of a direct connection between a documentary film and a social change is when the documentary's audience is the changemaker.

5.9 Conclusion

It is important to recognize the "digital divide" in this chapter. Not all filmmakers, audiences, or changemakers have the same access to digital technology. In the early days of the documentary, films and projectors were brought to remote locations to reach the people who did not have access to this new technology. The same access to digital technology needs to be considered to reach those communities with limited computer and wireless networking resources. These communities need to be reached not only to deliver documentary film messages, but to provide a portal to those communities so their voices can be heard by the changemaker as well. Any remediation of the documentary film in the digital landscape needs to be inclusive of all communities and voices to be truly democratized and representative.

The documentary film is still a familiar entity in this new digital ecosystem of interactive and collaborative engagement. For the most part, it is still a linear format telling a factual story with the intent of engaging an audience emotionally and provoking it to action. The segments of evolution in form we have examined in this chapter pave the way for a grander remediation of the genre in ways that make the traditional linear documentary altogether unfamiliar and, as I will argue, significantly more effective in creating social change.

Bibliography

"About Us." *docsociety.org*. Web. Accessed August 24, 2018. https://docsociety.org.

Aufderheide, Patricia, and Jessica Clark. *Open Letter to FCC Commissioners*. Washington, DC: Center for Social Media, 2009. Print.

Chatoo, Caty Borum, and Angelica Das. *Assessing the Social Impact of Issues-Focused Documentaries: Research Methods & Future Considerations*. Washington, DC: Centre for Media and Social Impact, 2014. Print.

Dovey, Jon. "Documentary Ecosystems: Collaboration and Exploitation." In *New Documentary Ecologies: Emerging Platforms, Practices and Discourses*, ed. Kate Nash, Craig Hight, and Catherine Summerhayes. New York: Palgrave Macmillan, 2014. Print.

"Feds Release Details on Plans to Restore Prison Farm Program at Joyceville and Collins Bay Institutions." *CBC News*, June 21, 2018. Web. Accessed August 24, 2018. https://www.cbc.ca/news/canada/ottawa/kingston-prison-farm-cows-goats-1.4716232.

Ferreira, George, Ramirez Ricardo, and Allan Lauzon. "Influencing Government Decision Makers through Facilitative Communication via Community-Produced Videos: The Case of Remote Aboriginal Communities in Northwestern Ontario, Canada." *Guelph: Journal of Rural and Community Development*, 2009. Print.

Gaudenzi, Sandra. "Strategies of Participation: The Who, What and When of Collaborative Documentaries." In *New Documentary Ecologies: Emerging Platforms, Practices and Discourses*, ed. Kate Nash, Craig Hight, and Catherine Summerhayes. Basingstoke: Palgrave Macmillan, 2014. Print.

———. "User Experience Versus Author Experience." In *i-docs: The Evolving Practices of Interactive Documentary*, ed. Judith Aston, Sandra Gaudenzi, and Mandy Rose. New York: Columbia University Press, 2017. Print.

Gray, Jeff. "Canadian Documentary Warns Tax Havens Threaten Democracy." *The Globe and Mail*. Toronto: March 25, 2017. Web. Accessed February 21, 2018. https://www.theglobeandmail.com/report-on-business/industry-news/the-law-page/canadian-documentary-warns-tax-havens-threaten-democracy/article23409274.

Karlin, Beth, and John S. Johnson. "Measuring Impact: The Importance of Evaluation for Documentary Film Campaigns." *M/C Journal* 14, no. 6 (2011). Print.

Klause, Olivia. "Plot: Sin by Silence." *imdb.com*. 2009. Web. Accessed May 3, 2018. https://www.imdb.com/title/tt1074652/plotsummary.

Kopp, Ingrid. "Interactive Media Impact Working Group." Tribeca Film Institute, sandbox.tribecafilminstitute.org. Web. Accessed February 2016. http://sandbox.tribecafilminstitute.org/impact.

Manovich, Lev. *The Language of New Media*. Cambridge, MA: Massachusetts Institute of Technology, 2001. Print.

Marcus, Daniel. "Documentary and Video Activism." In *Contemporary Documentary*, ed. Daniel Marcus and Selmin Kara. New York: Routledge, 2016. Print.

McIntosh, Heather. "The Committed Documentary and Contemporary Distribution: A Look at Sin by Silence." In *Documenting Gendered Violence: Representations, Collaborations, and Movements*, ed. Lisa M. Cuklanz and Heather McIntosh. New York and London: Bloomsbury Academic, 2015. Print.

McLagan, Meg. "Imagining Impact: Documentary Film and the Production of Political Effects." In *Sensible Politics: The Visual Culture of Nongovernmental Politics*, ed. Meg McLagan and Yates McKee. New York: Zone Books, 2012. Print.

Nash, Kate. "Clicking on the World: Documentary Representation and Interactivity." In *New Documentary Ecologies: Emerging Platforms, Practices and Discourses*, ed. Kate Nash, Craig Hight, and Catherine Summerhayes. Basingstoke: Palgrave Macmillan, 2014a. Print.

———. "An Interview with Florian Thalhoffer, Media Artist and Documentary Maker." In *New Documentary Ecologies: Emerging Platforms, Practices and Discourses*, ed. Kate Nash, Craig Hight, and Catherine Summerhayes. Basingstoke: Palgrave Macmillan, 2014b. Print.

Nichols, Bill. *Blurred Boundaries: Questions of Meaning in Contemporary Culture*. Bloomington and Indianapolis: Indiana University Press, 1994. Print.

Nime, Nicole Marie. *The Impact of Digital Technology on Documentary Distribution*. London: University of London, 2012. Print.

Nisbet, Michael C., and Patricia Aufderheide. "Documentary Film: Towards a Research Agenda on Forms, Functions, and Impacts." *Mass Communication & Society* 12, no. 4 (October–December, 2009). Print.

Perlmutter, Tom. "The Interactive Documentary: A Transformative Art Form." *Web: Policy Options*. Web. Accessed February 21, 2018. http://policyoptions.irpp.org/magazines/policyflix/the-interactive-documentary-a-transformative-art-form.

Rosenthal, Alan, and John Corner. "Introduction." In *New Challenges for Documentary* (2nd ed.). Manchester and New York: Manchester University Press, 2005. Print.

Spence, Louise, and Vinicius Navarro. *Crafting Truth: Documentary Form and Meaning*. New Brunswick, NJ and London: Rutgers University Press, 2011. Print.

Varner, Gary Matthew. "*Koyaanisqatsi* and the Posthuman Aesthetics of a Mechanical Stare." *Film Criticism* 41, no. 1 (February 2017). Web. Accessed July 30, 2019. https://doi.org/10.3998/fc.13761232.0041.104.

Verellen, Emily. "From Distribution to Audience Engagement: Social Change Through Film." *The Fledgling Fund* (August 2010). Web. Accessed May 10, 2017. http://www.thefledglingfund.org/impact/From%20Distribution%20to%20Audience%20Engagment.pdf.

Whiteman, David. "The Evolving Impact of Documentary Film: Sacrifice and the Rise of Issue-Centered Outreach." *Questia*. Chicago: Cengage Learning, 2007. Print.

Winston, Brian. "Foreword." In *i-docs: The Evolving Practices of Interactive Documentary*, ed. Judith Aston, Sandra Gaudenzi, and Mandy Rose. London and New York: Wallflower Press, 2017. Print.

Winton, Ezra, and Svetla Turnin, ed. *Screening Truth to Power: A Reader on Documentary Activism*. Montreal: Cinema Politica, 2014. Print.

Filmography

'Til the Cows Come Home, Director, Lenny Epstein. Telltales Media and Milkpail Productions, 2014.

Blackfish, Director, Gabriela Cowperthwaite. Manny O Productions, 2013.

Sin by Silence, Director, Olivia Klaus. Orange, CA: Quiet Little Place Productions, 2009.

The Corporation, Director, Harold Crooks. Big Picture Media Corporation. 2003.

The Price We Pay, Director, Harold Crooks. InformAction Films, 2015.

Urge Gov Brown to Pass AB 593 & AB 1593. n.d. SinBySilenceDoc, 2012. https://www.youtube.com/watch?v=sp3dbPziiu8&t=2s.

VTR St-Jacques, Director, Bonnie Sherr Klein. National Film Board of Canada, 1969.

CHAPTER 6

Visible Volume: The Multilinear and Database Documentary

6.1 Introduction

As we have examined in the previous chapter, the digital landscape offers many technical advances in creating documentary projects, mobilizing social activism, and offering new methods and approaches to combine the two to achieve the elusive goal of influencing social change with non-fiction film. This chapter examines the intersection of the documentary and the digital in two experimental areas to determine how successful this specific remediation is in communicating, informing, and influencing the changemaker: the multilinear and database documentary.

These two approaches relate to theory we examined in previous chapters concerning ecocinema and the digital, interactive documentary. As a design methodology, the multilinear documentary structure mirrors the Internet. With a multitude of websites available to visit at the click of a button on the World Wide Web, so too are similar selections being made available in the construct of a multilinear documentary platform. Several film fragments, usually related to the documentary's subject matter in some way, can be viewed by audiences by the same click of a button. The choices of what to watch and when to watch are now in the hands of the viewer, instead of the documentary's editor and director. This non-narrative approach, however, is not without structure, as David Bordwell and Kristin Thompson explain in their book *Film Art: An Introduction*:

M. Terry, *The Geo-Doc*, Palgrave Studies in Media and Environmental Communication,
https://doi.org/10.1007/978-3-030-32508-4_6

> a film is not simply a random bunch of elements. Like all artworks, a film has form. By film form in its broadest sense we mean the overall system of relations that we can perceive among the elements in the whole film. (Bordwell, Thompson, 57)

Even if the elements of the documentary film seem randomly placed in a multilinear format, there still exists a "system of relations" to the fragments that the audience employs through their choices of selection.

The other new way of presenting and consuming the digital documentary in its multilinear structure relates to the spatial analysis this affords. Regarding the film fragments of certain locative documentary, subjects can provide a semiotic storyline much like the implicit narratives being used in ecocinema. Digital media theorist Adrian Miles identifies the sweeping transformative properties that impact the documentary when remediated in this new form:

> Nonfiction relies on the idea of an authentic relation to the world. As a consequence documentary makers and audiences want the work to be 'truthful'. This is most simply realised through decisions about what to film, what to keep, what order to edit these into and whose voice appears when, and with what visuals. However, in multilinear online documentary each of these things (and all) can change. (Miles, "Authenticity," 2014)

6.2 The Multilinear Documentary

As we examined in the previous chapter, the digital domain affords new tools of production, exhibition, and distribution to the documentary filmmaker, but are these tools beneficial advancements to the filmmaking craft and the documentary as an influential communications tool or do they complicate and thereby hinder the documentary experience for the audience? The technology of the craft may be improving, but at what cost to its audiences and its activist goals?

The interactive documentary is redefining its audiences as users in the digital domain as more and more engagement tools develop in the online production, exhibition, and dissemination of these films. While audiences/users are being engaged more than ever before, according to Kate Nash in her essay "Clicking on the World: Documentary Representation and Interactivity," "there is no necessary connection between interactivity and audience empowerment" (Nash, 50). As we examined previously, she sets a framework for examining the interactive documentary conceived as multidimensional "in which the actions of users, documentary makers, subjects and technical systems together constitute a dynamic ecosystem" (Nash, 51).

Within this framework, we can examine the production and engagement of the multilinear documentary format. It is important at this point to distinguish the difference between the terms "interactive" and "multilinear" as they pertain to the digital documentary. They are often seen to be synonymous, but, in fact, are quite dissimilar. While all multilinear documentaries are interactive—they have to be in order to engage in them—not all interactive documentaries are multilinear. A YouTube video, for example, is linear, but its presentation platform affords the viewer interactivity by leaving comments and sharing. Furthermore, multilinear documentary film projects are not necessarily non-linear. The overall project offers the user a selection of film fragments to view in any order they choose. This structure appears non-linear, yet each film unit, when made as an individual linear documentary film, is not non-linear. Collectively, a project of multiple linear documentary film units constitutes a multilinear film project. Conceptually, the idea of a multilinear film is nothing new. We have seen examples of non-interactive, multilinear film viewing in popular culture, going back as far as the 1960s: the video surveillance wall in Albert Finney's *Charlie Bubbles* (1967), for example. At the film's 00:16:22 mark, the film prophetically displays scenes of non-fiction in the various rooms, hallways and, indeed, all spaces of Charlie's mansion in a surveillance-like manner, well before the technology was common in family homes or even in corporate security offices.[1] Successful British writer Charlie Bubbles (Albert Finney) watches multiple screens of live video of rooms, hallways, virtually all spaces in his home simultaneously, surveying his world in real time. A review in *The New York Times* describes this scene this way: "One imagines that God observes his own universe by some similar arrangement" (Alder 1968). The suggestion here is that the multilinear format of the documentary film may grant its audiences an

[1] Surveillance technology goes back as far as 1942 when German engineer Walter Bruch invented the closed-circuit television camera to watch Nazi rocket launches from a safe distance. The first documented use of multiple cameras and monitors for surveillance was in 1968 by the city of Orleans, New York, as a means of fighting crime (Lee, 212).

omniscient comprehension of its profiled social issue in a way usually ascribed to deities.

During that same year, Expo 67 in Montreal unveiled a multilinear project installation on a grand scale called *Labyrinth*, produced by the National Film Board of Canada under the direction of Colin Low and Roman Kroitor ("Labyrinth," *http://cinemaexpo67.ca/labyrinth*). The non-interactive, multiscreen project incorporated screens fifty feet high, viewed by audiences in stacked balconies. In a prophetic analysis of his own work, multiscreen editor Tom Daly said "If you are going to control people's choice, what is the value of multiscreen? Why not let them make their own choice?" (Feldman, 42). More than thirty years later, this very idea of interactivity was added to the concept of the multilinear documentary, thanks to the affordances provided by digital technology.

Referenced earlier, one of the leading scholars in multilinear theory was Adrian Miles of the Royal Melbourne Institute of Technology who contributed a major body of work to the study of multilinear documentary storytelling. His first encounter with it was in 1991 when he remarked that it "presented itself as an overwhelming paradigm shift in the practice of cinema studies" (Miles, *Soft Cinematic Hypertext*, 18).

In addressing the unique affordances of the multilinear format in film, Miles explains its relational properties through an example of a man being shot.

> As an example, consider the very simple case of a multilinear video consisting of three, thirty-second clips. One sequence is a close-up of a hand holding a gun and the trigger being squeezed. The second, a body falling to the floor. The third, a figure entering a room and picking up a gun. These are placed in a multilinear work where any can occur at any point in the sequence. (Miles, *Soft Cinematic Hypertext*, 67)

Each of these clips, he argues, can be played in any order, and with each sequential selection, a different narrative is presented. If the third scene is selected to start the sequence, the story may reveal the killer. If that scene comes last, the person picking up the gun could be someone who heard the shot, discovered the body, and wants to examine the murder weapon. Miles calculates that if each of these three scenes is thirty seconds long, the linear approach of watching each one in order yields a ninety-second film, but with the multilinear affordance of scene selection or choice, there are various combinations: "given the number of possible combinations of the three clips… there are eighteen combinatory possibilities (three x two x three) and now the work is nine minutes long if you wanted to see all possible combinations" (Miles, *Soft Cinematic Hypertext*, 67). He argues here that documentary fits this format better than fiction since many of the scenes in a non-fiction narrative are not reliant on scenes that come before or after. "In documentary you can easily join any image with any other image, continuity of action or narrative is not a formal requirement, the didactic nature of the work largely carrying what might otherwise be thought as narrative leaps" (Miles, *Soft Cinematic Hypertext*, 21). In a multilinear project, scenes related to the social issue being presented do not have to relate to each other to advance the narrative. Database documentary projects, in particular, present a wealth of images that can be viewed in any order and still deliver the film's overall message. For example, in Ruttmann's *Berlin: Symphony of a Great City* (1927), any of the shots of Berlin can be viewed in any order to appreciate the message in the title: that Berlin is a great city. A more current example is Jeff Orlowski's documentary of glacier collapse, *Chasing Ice* (2012). The impressive time-lapse photography of melting glaciers shot in Greenland and Alaska can be viewed in any order to understand that these frozen rivers are disappearing at an unprecedented and alarming rate.

Had these films been re-presented in a multilinear format, audiences could have made the choices Daly called for in 1967, and with these selections they would create relations with the film fragments that come before and after, possibly revealing an implicit narrative with this user-generated relational construct. By selecting the parts, many different sums of the whole can be realized: "(I)n their combination a context and comparison is established that generates significance between the otherwise disparate (film clips)" (Miles, *Soft Cinematic Hypertext*, 36).

One must be careful, warns communications professor Zoë Druick, of building a multilinear project that puts "collection over selection" (Druick 2018) or quantity over quality. This approach renders the project difficult to navigate and includes film fragments that may lack the content or production quality necessary to contribute significantly or meaningfully to the project.

In his essay "Database as a Symbolic Form," new media theorist Lev Manovich argues that there are two dimensions at play in a multilinear documentary that take the place of a traditional narrative in a linear documentary: the syntagmatic and the paradigmatic (Manovich 1999). The former represents the elements or fragments of the projects, that is, the film units. These represent an explicit narrative. The latter represents the user's imagined processing of these images. This is the implicit narrative. Roland Barthes explains this relationship best in his book *The Elements of Semiology*: "The units which have something in common are associated in theory and thus form groups within which various relationships can be found" (Barthes, 58).

One of the first software programs developed specifically to create interactive multilinear film projects was launched in the late 1990s by Florian Thalhofer as a documentary film project about alcoholism, comprising part of his master's thesis at Berlin's University of the Arts. The film he made was on the subject of the Korsakoff Syndrome—an amnestic disorder caused by thiamine deficiency—and the software he created to make this film was given the same name, but with a different spelling: the Korsakow system ("Korsakow," *http://www.thalhofer.com*).

The experimental film project depicting multiple film clips or media files (Thalhofer calls these Smallest Narrative Units or SNUs) on the single platform of a computer screen caught the attention of Matt Soar, a professor at Montreal's Concordia University who had just launched the Concordia Interactive Narrative Experimentation and Research Group (CINER-G, pronounced *synergy*) with Monika Kin Gagnon in 2000 (Sarkissian 2010). At the time, Thalhofer had taken his multilinear experiment as far as he could: "I didn't know about filmmaking," said Thalhofer. "I didn't know how to properly build stories. I did everything wrong. By doing everything wrong, I invented a new path" (Sarkissian 2010). After being recruited by Soar and his CINER-G group, Korsakow went through a major overhaul. The system was refined and relaunched in 2009, with a goal of making the software accessible to everyone.

> "I'm always on the lookout for new ideas and new platforms for creative expression," says Soar. "There's nothing quite like Korsakow out there. With a weekend and a modicum of skill, anyone can make a K-film. It's about accessibility." (Sarkissian 2010)

As examined in the previous chapter, the democratizing of the documentary filmmaking process is at the heart of the collaborative component of the interactive documentary. Extending from the participatory mode, so richly employed by Ivens, Tessier, and Low to effect social reform for the people their films profiled, the multilinear software created by Thalhofer and developed by CINER-G now allows anyone to tell their own story in this unique fashion. Just as the PortaPaks were put in the hands of community members in *VTR St-Jacques* to tell their own stories, Kosakow is now being put in the hands of anyone in the world with a computer and access to the Internet to tell their stories.

The concept of "too many options" is at the heart of the multilinear documentary. Supporting media such as additional video, web-links, texts, and pictures may provide an enhanced presentation of data related to the documentary's subject matter, but engaging in them may interrupt the traditional experience of simply "looking at and listening to" a linear documentary. Too often in a multilinear project, all media are presented simultaneously, requiring choices which on the one hand empower the user, while on the other hand establish an interface that could disrupt the flow of traditional audience engagement.

Adrian Miles argues that while the format of the multilinear documentary as conceived by the Korsakow system holds promise for the future of documentary storytelling, its current iteration is problematic:

> Korsakow foregrounds a computational rather than narrative logic, which is perhaps why it has been so misunderstood critically and why to work successfully with Korsakow requires some surrendering of your agency to the procedural demands of the unit or system…To make sense in and of Korsakow, for making and viewing, it needs to be recognised, deeply and seriously as a unit, system, and actor network. We are actants within this system, but never its centre or origin. For documentary this proposes a reading and making of the world that is not pre-determined nor fully controllable, for maker, reader, narrator, or the work. (Miles, 219)

In terms of a remediation of the documentary form, Miles suggests that multilinear creation systems like Korsakow are still in their infancy in terms of producing projects that both filmmakers and audiences are comfortable using. Miles suggests when the agency with which we usually engage in the production and viewing of documentary film is abandoned, so too is the traditional experience of film spectatorship, leading to misunderstanding at best, frustration with engaging with an unfamiliar interface at worst. But at the core of the multilinear format is a problem Miles identifies as the chief challenge in making these projects truly documentary: the storytelling structure. The traditional linear narrative is easy to follow as the road from beginning to end has been paved for us by the filmmaker and his editorial crew; however, in a multilinear format, there exists a "simple (but deeply sophisticated) problem of how things that are fragmentary wholes can be presented and related to each other in a way that enables the production of a new and comprehensible whole" (Miles, "Interactive Documentary," 71).

Echoing these concerns is digital humanist Siobhan O'Flynn. She argues that the multilinear format, and the Korsakow system in particular, suffers from a lack of editorial order, and the platform underuses the digital enhancements available to it in its residence online:

> (I)nteractive online films (documentary or drama) designed in this form are more often reified experiences that rarely create an emotional resonance with the interactant and this is a consequence of two factors. The first is that the removal of a fixed editorial structure results in the absence of a sense of a narrowing horizon of choice leading to a dramatic climax and conclusion…The second is the often 'flat' or 'static' interface design that takes only minimal advantage of the affordances of web interfaces, where the focus of interaction is only on what content to view next. (O'Flynn, 146)

Coming to its defence, Korsakow's co-creator Matt Soar suggests that "O'Flynn appears to be conflating her own expectations and tastes—specifically, an overriding investment in 'dynamic interface design' (O'Flynn, 147)—with the actual affordances of Korsakow as a tool for filmmaking" (Soar, 161). These critiques are helpful signposts for designing a multilinear project that addresses these concerns conceptually and in practice.

The creators of Korsakow, Thalhofer and Soar, define the individual film clips in their projects as "smallest narrative units" (SNUs) because it is their intention that these clips tell a story, hence the descriptor "narrative." Adrian Miles prefers to describe these SNUs as having "granularity." This term is loaded with descriptors of the clips such as duration and relationality to each other (Miles, "Programming Statements," 147). Traditionally, he argues, granularity in film theory refers to the individual shots that, when edited together in a specific order, form the linear narrative of a film. In multilinear filmmaking, the granules of the project are still the shot, but now the order of their viewing is controlled by the spectator, not an editor.

> (T)he granularity of the system is such that it can be subdivided in terms of duration and still be immanently meaningful… once we recognise the importance of such external relations we can see that any shot must, by defi-

> nition, exist in a multiple set of possible relations with other shots. (Miles, "Programming Statements," 148)

I believe this is where many multilinear projects fail. The *multiple set of possible relations with other shots* creates the opportunities for several "bad connections," relations that the filmmaker might have conceived, but that their audience may not make. It is therefore essential in the creation of a multilinear documentary to ensure that the SNUs, the film fragments, and their granularity are not dependent upon relating to each other through connectivity. How can this be achieved? By making the fragments autonomous without sacrificing their relationality.

I have created a multilinear documentary film project that we will explore in the next chapter that applies this theory. If the film units are independent "mini-docs" on a subject consistent with each other, they maintain relationality without requiring a specific sequence of viewing to provide context. They relate to each other by subject matter while standing on their own with independent narratives. What makes the multilinear project more robust than a linear documentary on the same subject are the various analyses that emerge when the viewer relates the film units with each other, as well as with spatial and temporal analysis affordances.

6.3 The Database Documentary

The documentary film has seen many remediations since entering the digital era: the i-doc, the web-doc, the eco-doc, the multilinear documentary, as we have examined in this and other chapters. In fact, one digital documentary scholar, Atalanti Dionysus, has identified as many as fourteen versions of the documentary remediated by the web (Dionysus 2018). What they all have in common is their capacity to be yet another version of the genre, the database documentary.

Many scholars—Manovich (2001), Lovink (2008), Keep (2015)—argue that all digital documentary projects, to some degree, are database documentaries. The digital domain in which they reside can accommodate infinite content provided by filmmaker, audience, and even community participants. As such, database documentaries are often referred to as "living documentaries" because of their capacity to continually grow. As Sandra Gaudenzi so accurately defines it in her book, *The Living Documentary: From Representing Reality to Co-creating Reality in Digital Interactive Documentary*, a living documentary is "an assemblage composed by…elements that are linked through modalities of interaction… and can be more or less open to transformation" (Gaudenzi, 84).

A living documentary, simply put, is a film without end, and this fits quite comfortably in the multilinear format that affords temporal and spatial analyses. These considerations provide the continually growing database with content that transforms both the explicit and the implicit narratives of the story. In the case of a social issue documentary, using a database documentary project as a policy resource, corresponding social reform can take place not just once—as in a linear documentary—but several times as new, updated data become available within the project.

In his article "The Art of Watching Databases," Geert Lovink, founding director of the Institute of Network Cultures, argues that "(w)e no longer watch films or TV; we watch databases" (Lovink, 9). In it, he explains that our encounters with database projects are preferred to linear projects based on their ability to be searched. "Which search terms will yield the best fragments" (Lovink, 9), he asks. Since we now have the choice Daly called for in 1967, we tend to stay engaged with a database documentary project much longer than the ninety minutes we used to give to a traditional, linear documentary film (Lovink, 12).

Another definition of the database documentary is provided by Anne Burdick et al. in the book *Digital Humanities*:

> Database documentaries are multi-linear. They are not watched, but rather performed by a reader/viewer who is provided with a series of guided paths; and, unlike the cinematic documentary, which is free-standing, database documentaries may be built on multiple, overlapping databases. (Burdick et al., 54)

But can these seemingly random collections of images—film fragments—presented in a multilinear space be understood as narratives? Can they tell a story if they are viewed out of context without any ordered sequence? Manovich argues that they can, by presenting a case using the example of Vertov's *Man with a Movie Camera*—a non-digital project made in 1927. Manovich describes Vertov as the "major database documentary filmmaker of the twentieth century" (Manovich, 239) and that "*Man with a Movie Camera* is perhaps the most important example of a database imagination in modern media art" (Manovich, 239). The main difference between this analogue film and its digital counterparts is the lack of audience interactivity, but even this is suggested through scenes showing the audience *in* the film, presumably "interacting" with the film they are seeing, the same one in which they appear. But how does this film serve as a database documentary, and does this structure yield a film that carries a message of social reform, and how effective is this technique in delivering this message to its audience? Manovich offers an explanation:

> In one of the key shots repeated few times in the film we see an editing room with a number of shelves used to keep and organize the shot material. The shelves are marked "machines," "club," "the movement of a city," "physical exercise," "an illusionist," and so on. This is the database of the recorded material. The editor—Vertov's wife, Elizaveta Svilova—is shown working with this database: retrieving some reels, returning used reels, adding new ones. (Manovich, 239–240)

The way the film is cut, Manovich argues, results in the film's narrative: "*Man with a Movie Camera* traverses its database in a particular order to construct an argument. Records drawn from a database and arranged in a particular order become a picture of modern life…Its subject is the film-maker's struggle to reveal (social) structure among the multitude of observed phenomena" (Manovich, 240). Database documentaries, therefore, possess the ability to reveal implicit narratives necessary in presenting arguments for social reform.

One of the scholarly affordances of the digital world is the way multiple elements can be analysed and investigated spatially. The Spatial Humanities offers us new ways of investigating interdisciplinary scholarship from a perspective of space. In examining the documentary film through this lens, we see relational patterns emerge, both spatially and temporally, providing data and narratives not necessarily evident in the documentary film itself.

6.4 The Spatial and Temporal Turns

The digital world has afforded us another method for interdisciplinary study: space. In their introduction to the book, *The Spatial Humanities: GIS and the Future of Humanities Scholarship*, editors David J. Bodenhamer, John Corrigan, and Trevor M. Harris explain that humanists have long been familiar with space as a "concept and metaphor," giving as examples "gendered space," "the body as space," and "racialized space," but now scholars are reviving their interest in "the influence of physical or geographical space on human behaviour and cultural development" (Bodenhamer et al., vii).

A lot of this renewed interest, they argue, has come due to the rise in popularity and use of a digital tool known as Geographic Information Systems (GIS) mapping. This online analytical programme "lies at the heart" of the spatial turn that "uses location to integrate and visualize information" (Bodenhamer et al., vii). In turn, this allows for analysis of data in large volumes to be seen from the metaphoric and arguably actual height of 10,000 feet. This unique perspective in scholarship facilitates the discovery of new data or, as I have been arguing in its application to documentary film, implicit narratives. "Within a GIS users can discover relationships that make a complex world more immediately understandable by visually detecting spatial patterns that remain hidden in texts" (Bodenhamer et al., vii).

Within the construct of a GIS map, temporal analysis can now be added, providing an additional discipline of study. Specifically, history can be applied here. Decorated American humanist Edward L. Ayers believes that maps and history are "deeply complementary when looked at as landscapes and timescapes." Invoking the work of theorist Denis Cosgrove, Ayers argues that "both provide a way of reversing divisibility, of retrieving unity, of recapturing the sense of the whole. Maps and histories do the same kind of work in different disciplines, in different dimensions of human experience" (Ayers, 3).

This research binary is a valuable asset to the social issue documentary, especially in a multilinear format and presented on a platform of a GIS map. It contextualizes the profiled issue in new ways, affording the changemaker a deeper understanding of the issue through new spatial-temporal knowledge not usually available in traditional, linear documentaries. This facilitates the creation of more accurate policy to improve conditions for the disenfranchised members of the profiled community.

To demonstrate a GIS platform for a multilinear documentary film project, a sample has been created, called *The Documentary Film World*[2] (see Fig. 6.1). The project showcases most of the documentary films referenced in this monograph in their entirety, as well as other high-profile and award-winning social issue documentaries, situating them at the geographic coordinates of where each film is set and listing the year of produc-

[2] The project title is hyperlinked, but it can also be accessed at this address: https://www.google.com/maps/d/viewer?mid=1dJWzcyud63ntYDr9nt5sfobJ4v-neE6y&usp=sharing.

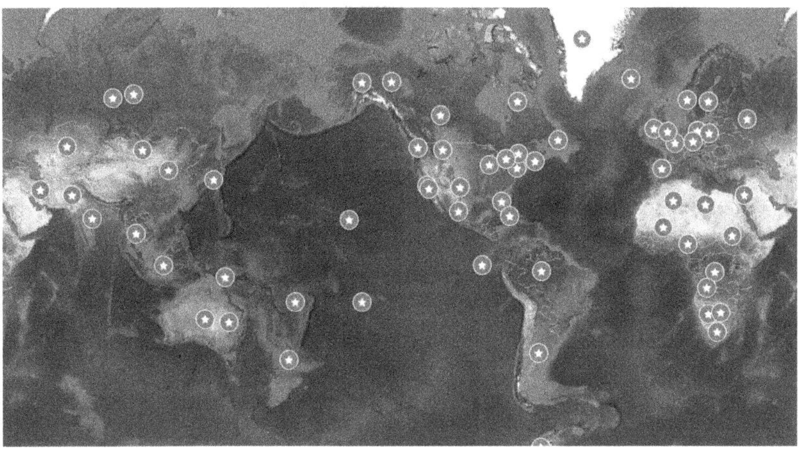

Fig. 6.1 *The Documentary Film World*, creator, Mark Terry, 2018. https://www.google.com/maps/d/u/0/edit?mid=1dJWzcyud63ntYDr9nt5sfobJ4v-neE6y&ll=2.257225636291693%2C0&z=2

tion in its metadata. This design structure affords the user the opportunity to analyse the data (the films), both spatially and temporally. For example, there are two documentary films located in the South Pacific island nations of Tahiti, *Tabu: The Story of the South Seas* (1930), and Fiji, *Among the Cannibal Isles of the South Pacific* (1918). The subject of each documentary is an ethnographic examination of indigenous people living in remote parts of the world, specifically the islands of the South Pacific. A temporal analysis reveals that these films were made in 1918 and 1930, the early days of documentary filmmaking. With this very small sample set of two, a user may conclude that ethnographic study in this part of the world was a popular subject to profile for documentary's early filmmakers. Further analysis may reveal that the privileged filmmakers from the West might represent some strategic colonial interest in these island nations. In fact, an extended temporal analysis of films posted in this database and made prior to 1930 will reveal two other documentary films of remote locations (the Arctic), showcasing similar ethnographic studies of its native people: *Nanook of the North* (1922) and *Romance of the Far Fur Country* (1920). As two, or even four, films are not enough of a data set for any conclusive analysis, it may prompt a deeper investigation into the ambitions of colonial and ethnographic filmmaking in the early days of the documentary, a theory not as easily conceived without the aid of this data visualization tool.

A project such as this also serves as a living database documentary—a digital library of social issue documentary films made worldwide from the late nineteenth century to the present and into the future. It also affords the curation capacity for community involvement and representation. It is not a single-authored work, like a linear documentary often is, but one that can include all voices in the social issue documentary. As Roland Barthes might say, this technology may very well signal the death of the director. The director of a multilinear project does not even have to shoot one frame of film. Serving more as a digital architect of film units produced by others, the multilinear director (perhaps more accurately redefined as a curator) creates a digital platform for the exhibition of film units made by others.

The Documentary Film World is an example of one such project. It was created with a beta software developed by Google known as *Fusion Tables.*[3] Launched in June 2009, Fusion Tables offered users an open-source digital playground to experiment with the limited affordances offered by this new Geographic Information Systems technology. In a paper written by Fusion Tables' developers, they explain why they designed the GIS software the way they did:

> The goal of Fusion Tables is not to replace traditional database management systems and applications, and neither is the goal to simply move such applications into the cloud. In contrast, the objective is to offer data management functionality that exploits today's computing environment in order to effectively enable new users and uses of data management technology. (Gonzalez et al., 1062)

Exploiting today's "computing environment" addresses O'Flynn's criticism of the Korsakow system in limiting its accessibility to "the affordances of web interfaces" (O'Flynn, 146). Its open-source, no-cost access, and relatively user-friendly interface democratizes the technology previously privileged only to those with an understanding of coding languages. In much the same way the PortaPak, the Internet, mobile devices, and Korsakow put documentary filmmaking into the hands of global communities, and not just the previously privileged professional filmmaker, Fusion

[3] In December 2018, Google announced it will be taking the experimental GIS software Fusion Tables offline as of December 3, 2019. Its recommended replacement is Google's *My Maps.* Both GIS projects created by the author—*The Documentary Film World* and *The Youth Climate Report GIS Project* (Chap. 7)—have been migrated to *My Maps.*

Tables, and other GIS software like it introduce another democratizing technology for documentary film production to people worldwide. Google has decided to take Fusion Tables offline and now offers *My Maps*[4] as an open-source alternative. The software is more robust in terms of interface, colour palettes, and interactive options.

6.5 Quantitative GIS

Scholars such as Ian Gregory, a professor of Digital Humanities at Lancaster University, Allister Geddes, a lecturer in Human Geography at the School of Social and Environmental Sciences, and Paul S. Ell, founding director of the Centre for Data Digitisation and Analysis at Belfast's Queen's University, are suspect of the GIS platform as a valuable study tool for humanists. In their book, *Toward Spatial Humanities*, Gregory and Geddes point out that the early use of GIS was controversial: "Opponents argued that it marked a lurch toward an unacceptable form of positivism with no epistemology or treatment of ethical or political issues" (Gregory, Geddes, ix). This is true when GIS is regarded in its natural state, but through multilinear documentary films populating geographic locations on a GIS platform, the ethical and political nature of documentary invigorates the GIS platform to provide that which its critics claim is missing.

Another dissenting voice is expressed by Ell who, arguing that Humanities GIS is still in an "embryonic state," suggests that the quantitative nature of GIS mapping is of no interest to humanists. "GIS…tends to be limited to certain disciplines, such as Historical Geography, is largely quantitatively based, analyzing census data, for example, and focuses mainly on maps and other visualizations. Humanities GIS conceived in these terms will always have limited appeal" (Ell, 144). The challenge, therefore, is not to use GIS technologies simply in these ways, but to develop dynamic, new approaches of introducing ethical and political content in epistemological structures. As this is in the nature of the social issue documentary, I argue that geographically situated documentary film fragments can expand the GIS infrastructure to include this humanities perspective. Furthermore, if the participatory nature of community representation in the films is added, as well as global, interactive collabora-

[4] The geo-doc projects created for and examined in this book have been migrated from Fusion Tables to *My Maps*.

tion afforded by its digital domain, we can now construct a platform that provides multiple disciplines of study and a more robust instrument of investigation for humanists, social scientists, and policymakers who work in these areas. Ell concedes that while not yet practised, GIS technologies hold the possibility for this kind of structure and resulting research.

> All humanities research sources contain key elements that can be used by a GIS. First, information has a spatial location whether expressed precisely or generally. Second, all humanities data contains a temporal marker even though its granularity may vary from minutes to a decade, century, or more…Third, all humanities data can be classified by subject…GIS is able to treat date and location as an attribute and as such allows for more exacting organization of information than the traditional subject-based approach. This functionality, coupled with the proliferation of electronic resources, offers new opportunities for resource discovery and use by humanities scholars. (Ell, 145)

This nascent technology in the arena of research provides epistemological precepts upon which traditional disciplines in the humanities can be added to build an infrastructure of new study practices that yield new knowledge. This emboldens the quantitative nature of GIS with the qualitative nature required by policymakers engaged in social reform projects. Consequently, a remediation of the documentary film within this framework may enhance its ability to effect social change.

6.6 Qualitative GIS

If we consider a GIS platform as being inherently quantitative in terms of measuring time and spatial distance between its situated data, then the technology is one-dimensional in its disciplinary approach to study. However, when we introduce content that represents data in more interdisciplinary forms, including such perspectives as ethics, morals, social studies, art, anthropology, gender, ethnography, and globalization, just to name a few, a qualitative nature emerges. Naturally, all of these disciplines can be represented by film, and film can populate GIS platforms as SNUs. More specifically, documentary film, containing both quantitative and qualitative data, can provide a well-rounded resource of factual information, with humanistic frames of reference viewed through the lenses of spatial and temporal analyses.

Using GIS in this manner is beginning to attract the activist in building stronger arguments for change, one that combines dispassionate data with impassioned social justice arguments in an easily accessible and widely distributable digital format. One experiment in this approach comes from climate change journalist and activist Mason Inman who created a GIS map of fracking sites across America called *Fracking in the North Dakota Oil Boom* (2015).[5] In a report on the contentious issue of fracking—the process of injecting liquid at high pressure into subterranean rocks, boreholes, and so on, so as to force open existing fissures and extract oil or gas[6]—Inman created a map of all North Dakota fracking projects to provide transparency for the general public in environmentally threatening practices that are not widely publicized. "I haven't seen any interactive maps out there that let ordinary citizens easily explore the extent of this boom" (Inman, "Adventures in Mapmaking," 2015). While admitting not to know what to do in activist terms with this data visualization project, Inman says he created it to assist those who do. "This map alone won't answer those big questions I had, about how long the boom in North Dakota might last. But tracking how the boom is unfolding is one crucial part of understanding the potential for the long-term" (Inman, *Adventures in Mapmaking*, 2015).

GIS scholar and geography professor at the City University of New York, Marianna Pavlovskaya cites this project as an example of qualitative mapping recruited for activist intentions using terms more commonly referenced in descriptions of social issue documentary theory: "Qualitative GIS has much to contribute to neogeography in terms of understanding the collective practices of geographic knowledge production. This research on community participatory mapping and public participation GIS commonly involves academics who mediate the interaction between communities, GIS technologies, and policymakers" (Pavlovskaya, 7).

With terms such "participatory mapping" used in connection with "communities" and "policymakers," we see the influence of the social issue documentary in its courtship of GIS mapping technology as an influ-

[5] Inman, Mason. *Fracking in the North Dakota Oil Boom*, 2015. Accessed: April 29, 2018. https://api.tiles.mapbox.com/v4/wiredmaplab.khoba8jj/page.html?access_token=pk.eyJ1Ijoid2lyZWRtYXBsYWIiLCJhIjoiVXNSbEtxSSJ9.zTF03t8ogjPIEAouNzT4-Q#10/47.9554/-102.8815. Web.

[6] Oxford Dictionary. Accessed April 29, 2018. https://en.oxforddictionaries.com/definition/fracking. Web.

ential communications tool. How, then, can the documentary film be married to the digital platform of GIS?

Gregory and Geddes believe this is possible. "The first challenge is simply to create effective spatial databases of qualitative sources. The development of attributable data in a range of non-traditional formats such as still images, movies (read *documentary films*), and sound will definitely continue within and beyond GIS…Developing, extending, and disseminating these resources has the potential to pay high dividends for humanities GIS" (Gregory, Geddes, 175).

GIS theorist and author of the essay "The Geography of Film Production in Italy: A Spatial Analysis Using GIS," Elisa Ravazzoli agrees that film can play an important role in new knowledge mining through GIS technologies. Her analysis echoes the semiotic storytelling practices of ecocinema: "GIS…has the capacity to integrate and visualize different types of information, to detect spatial patterns, and to disclose relationships hidden in texts…spatial analysis can be used for the investigation of many aspects related to film. It enables the study of the intrinsic spatial relations between the elements within the film, (and) the analysis of the relationship between the space of the film and the real space" (Ravazzoli, 151–153).

A documentary project seeking to impact a social issue can theoretically situate film fragments of the subject geospatially through a multilinear storytelling structure. A subject such as poverty, for example, can have documentary segments focus on the social issue as it presents itself in different cities, urban and rural environments, and countries. Analysis of these SNUs spatially could ostensibly reveal patterns related to public policy or socio-economic influence. Further analysis temporally—assuming the films are made over the course of several years—could illuminate these patterns even more in terms of what socio-political conditions were in place at each time.

This creative structure is an early model for how to build a social issue database documentary—a narrative order needs to emerge from the apparent chaos of unrelated film fragments. The criticism we examined in the multilinear format also demands some form of relational order to establish a perceptible narrative. To this end, I propose a multilinear, interactive, database documentary film project that maintains a consistent subject and theme in each of its film fragments. These SNUs are not pieces of a story, but stand-alone stories in and of themselves. They serve as "mini-docs," independent of each other, yet all contributing to tell a story collectively. This approach satisfies the "look and listen" struggle expressed by

Rosenthal and Corner. The consistent subject of each film fragment also satisfies Miles' concern for relationality between the film units. To address O'Flynn's concern about the flatness of the Korsakow interface and its lack of editorial structure, I propose presenting this multilinear, interactive, database documentary film project on a platform of a GIS map. Since each film unit is an independent story, editorial structure is built into each unit. Furthermore, since all film fragments are stories about the same subject, differing primarily by geographic location, they all relate to each other not only thematically, but spatially and temporally as well.

6.7 Conclusion

In the previous chapters, we have explored how the Canadian tradition of participatory and community filmmaking resonated with the changemaker in a positive and productive way. We also saw how ecocinema and digital tools and platforms provided affordances that enhance the goals of the documentary film in informing and influencing positive social change. The multilinear format—when constructed in a relational manner—takes these remediated advances and amplifies the data and its relevant interpretations to provide the changemaker with a more informed understanding of the social issue profiled.

The next chapter will examine how this new documentary remediation works to serve the changemaker by incorporating the historical precedents, theories, methodologies, and approaches we examined to have worked most effectively in the previous chapters. As a case study, this GIS multilinear documentary film project addresses the global issue of climate change and how it is being used by the policymakers of the United Nations as a data delivery system.

Bibliography

"About." *The Ushahidi Ecosystem*. www.ushahih.com. Web. Accessed August 17, 2018. https://www.ushahidi.com/about.

"Analysis of the United Nation's Youth Climate Report GIS Project." *youthclimatereport.org*, September, 2017. Web. Accessed May 19, 2018. http://youthclimatereport.org/news-cat/analysis-united-nations-youth-climate-report-gis-project-helen-hughes.

"Canadian Film to Be Shown During UN Climate Meeting." *Canadian Press / CTV News*, August 18, 2009. Web. Accessed May 13, 2018. https://www.

ctvnews.ca/canadian-film-to-be-shown-during-un-climate-meeting-1.426585.

"Decision Adopted by the Conference of Parties." *Cancun: United Nations Framework Convention on Climate Change*, March 15, 2011. Web. Accessed May 5, 2018. http://unfccc.int/resource/docs/2010/cop16/eng/07a01.pdf#page=4.

"Interview with Erik Solheim." *youthclimatereport.org*, April, 2018. Web. Accessed May 19, 2018. http://youthclimatereport.org/?p=2428&preview=true.

"Lights, Camera, Marrakech." *United Nations Climate Change*. Marrakech, Morocco, November 9, 2016. Web. Accessed May 14, 2017. https://unfccc.int/topics/education-and-outreach/events%2D%2Dmeetings/global-youth-video-competition/2016-global-youth-video-competition-on-climate-change.

"Performance Peacekeeping." *Keeping the Peace – The UN Department of Field Service's and Peacekeeping Operations Use of Ushahidi*. www.ushahih.com. Web. Accessed August 17, 2018. https://www.ushahidi.com/blog/2018/08/08/keeping-the-peace-the-un-department-of-field-services-and-peacekeeping-operations-use-of-ushahidi.

"Recommendation 97." *Report of the United Nations Conference on the Human Environment*. Stockholm: United Nations, 1972. Print.

"See Inspiring Climate Action." *Resource: United Nations Framework Convention on Climate Change*, October 19, 2016. Web. Accessed May 19, 2018. https://unfccc.int/news/see-inspiring-climate-action-entries-to-global-youth-video-competition.

"The Antarctica Challenge: A Global Warning." *Sioux Falls Scientists*, n.d. Web. Accessed May 13, 2018. http://www.siouxfallsscientists.com/the-antarctica-challenge.html.

Boon, Tim. *Films of Fact: A History of Science in Documentary Films and Television*. London: Wallflower, 2007. Print.

———. "Science, Society and Documentary." In *The Documentary Film Book*, ed. Brian Winston. London: British Film Institute, 2013. Print.

Hallam, Julia, and Les Roberts, eds. *Locating the Moving Image: New Approaches to Film and Place*. Bloomington, IN: Indiana University Press, 2014. Print.

Hogben, Lancelot. "The New Visual Culture." *Sight & Sound* 5 (Spring, 1936). Print.

Hughes, Helen. *Green Documentary: Environmental Documentary in the 21st Century*. Bristol, UK: Intellect, 2014. Print.

Lapenta, Francesco. "Geomedia: On Location-Based Media, the Changing Status of Collective Image Production and the Emergence of Social Navigation Systems." *Visual Studies* 26, no. 1 (2011): 14–24. Print.

McQuire, Scott. *Geomedia: Networked Cities and the Future of Public Space*. Cambridge: Polity Press, 2016. Print.

Miles, Adrian. "Programmatic Statements for a Facetted Videography." In *Video Vortex Reader: Responses to YouTube*, ed. Geert Lovink and Sabine Niederer. Amsterdam: Institute of Network Cultures and XS4All, 2008. Print.

Thielmann, Thomas. "'You Have Reached Your Destination!' Position, Positioning and Superpositioning of Space through Car Navigation Systems." *Social Geography* 2, no. 1 (2007): 63–75. Print.

UNEP Partners with Film Company for Climate Change Conference. United Nations Environment Programme Press Release. Cancun: United Nations Environment Programme, November 22, 2010. Web. Accessed September 7, 2016.

Winston, Brian, Gail Vanstone, and Wang Chi. "Giving Voice." In *The Act of Documenting: Documentary Film in the 21st Century*. New York: Bloomsbury Publishing Inc., 2017. Print.

Filmography

Among the Cannibal Isles of the South Pacific, Director, Martin E. Johnson. Martin Johnson Film Company, Inc., 1918.

Antarctica Challenge: A Global Warning, The, Director, Mark Terry. Polar Cap Productions, Inc., 2009.

Blackfish, Director, Gabriela Cowperthwaite. Manny O Productions, 2013.

Charlie Bubbles, Director, Albert Finney. Memorial Enterprises and Universal Pictures, 1967.

Polar Explorer, The, Director, Mark Terry. Polar Cap Productions II, Inc., 2010.

Price We Pay, The, Director, Harold Crooks. InformAction Films, 2015.

Sin by Silence, Director, Olivia Klaus. Orange, CA: Quiet Little Place Productions, 2009.

Youth Climate Report, The, Directors, Mark Terry (2011–2014), Ray Kocur (2015). Neko Harbour Entertainment, Inc., 2011–2015.

GIS Project

The Youth Climate Report GIS Project, Creator, Mark Terry. Bonn, Germany: United Nations Climate Change, 2018.

CHAPTER 7

The Geo-Doc: A Locative Approach to Remediating the Genre

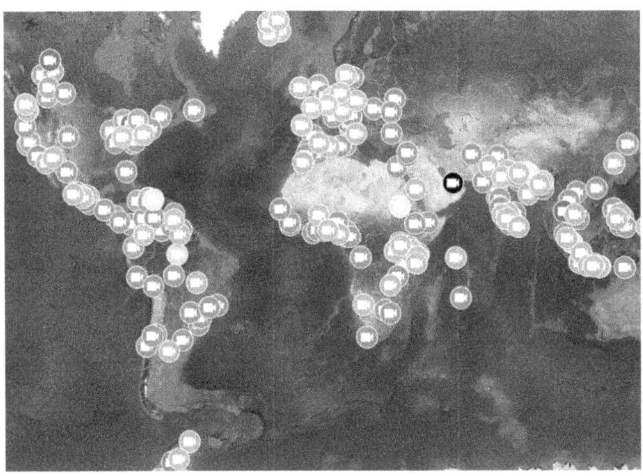

Fig. 7.1 *The Youth Climate Report GIS Project*, creator, Mark Terry, 2018. https://www.google.com/maps/d/u/0/edit?mid=13sPbdmhKOFINj9WsPllhmN_rAaHdhncV&ll=4.311127662008332%2C0&z=2

The Youth Climate Report GIS Project*, an interactive, multilinear, database documentary film project presented on a platform of a Geographic Information System map, currently used by the United Nations Climate Change communication's department as a data delivery system for delegates attending its annual climate summits known as the COP conferences.*

M. Terry, *The Geo-Doc*, Palgrave Studies in Media and Environmental Communication,
https://doi.org/10.1007/978-3-030-32508-4_7

7.1 Introduction

Geomedia, both as a term and as a study discipline, is relatively new as innovative locative technologies are being developed and used to analyse data geospatially, recruiting various media in this endeavour. This chapter investigates the role the documentary film is taking in geomedia as its digital remediation continues to evolve. Does this new platform for the social issue documentary enhance its ability to effect social change, or does it overcomplicate the audience's engagement with the film and its ideological messaging?

First coined by Thomas Thielmann in 2007 as a "genre" of various media incorporating location (Thielmann, 63), the term geomedia was first recognized as electronic applications of cartography, in particular car navigation systems. In examining this technology theoretically, Thielmann sees the genesis of a new source of knowledge:

> The dynamic representation of location in (car) navigation systems could therefore lead to a cross-disciplinary understanding of geography as a cognitive system in the widest sense of the phrase, which deals with the connections between subject and location. This not only re-evaluates the discursive positioning of the subject, but also integrates view points, difference strategies, contextualizations in space and time, and intelligible articulations. (Thielmann, 73)

The contexts of time and space provide unique affordances to documentary content used in geomedia projects. New temporal and spatial narratives exist which allow the opportunity for a greater understanding of the profiled documentary subject. Communications scholar Francesco Lapenta defines geomedia as existing media being used in this same epistemological way, but incorporating more of the affordances of the Internet:

> These technologies, that I call geomedia, are not new media per se, but platforms that merge existing technologies (electronic media + the Internet + location-based and Augmented Reality technologies) in a new mode of digital composite imaging, data association and socially maintained data exchange and communication. (Lapenta, 1)

The last two purposes begin to sound like the hallmarks of the social issue documentary film: community participation and new communications methods for speaking truth to power.

In his book on geomedia, *Geomedia: Networked Cities and the Future of Public Space*, Scott McQuire introduces the concept of interactivity as a defining feature:

> Geomedia is a concept that crystallizes at the intersection of four related trajectories: convergence, ubiquity, location-awareness and real-time feedback…What is different in the present is the way the distributed architecture of digital networks opens the potential for 'real-time' feedback from many-to-many supporting novel experiences of social simultaneity. (McQuire, 4)

These three definitions of geomedia situate the genre in its communicative nature alongside the social issue documentary film. Thielmann, Lapenta, and McQuire all refer to the ability of the technology to reach many and to have projects be created by many. As we examined in previous chapters, this echoes the successful methodology of the community-based and participatory documentary film.

The platform is fertile ground to grow the documentary in a newly conceived form. The documentary film has had brushes with geomedia in its past, primarily to advance storylines—a term collectively referred to as cinematic cartography. In their book *Locating the Moving Image: New Approaches to Film and Place*, Julia Hallam and Les Roberts argue that "cinematic cartography refers less to the presence of maps in film than to the cultural, perceptual, and cognitive processes that inform understandings of place and space" (Hallam, Roberts, 17–18). There are three main "shapes" of cinematic cartography: panorama, atlases, and aerial views which they refer to as "god's-eye" views (Hallam, Roberts, 18). This becomes significant when we recall *The New York Times* 1968 review of the film *Charlie Bubbles* when film critic Renata Alder describes the scene showcasing Charlie's multilinear surveillance wall as being suggestive of how "God observes his own universe" (Sect. 6.2). The panoptic view afforded by geomedia's specific construct of cinematic cartography suggests an omniscient understanding of that which is being studied, in this case, the issue presented in a multilinear format on a Geographic Information System platform. Cinematic topography in this form structures spatial knowledge: the documentary film fragments that comprise this geomedia platform tell their own stories in addition to implicit narratives that are revealed by relating the film units spatially and temporally, thereby granting the audience an "all-knowing" comprehension of the issue.

A geomedia technology that employs GIS mapping, but is not a documentary film project, is Ushahidi, a non-profit technology company whose name translates to "testimony" in Swahili. As we examined in Chap. 5, the digital divide can prevent those in remote communities from participating in social justice digital projects. Ushahidi's principal aim is to reduce that gap: "We design new products and initiatives with a global perspective. Our aim is to serve people with limited access in hard-to-reach places."[1] The crowdsourcing tool geolocates participants who upload data to contribute a collective voice on a social issue from their own mobile device. In peacekeeping efforts, the United Nations has employed Ushahidi since 2013. The UN's Department of Field Services uses Ushahidi to run its Situational Awareness and Geospatial (SAGE) programme. It is used to keep all field offices in Mali, Haiti, South Sudan, Lebanon, and the Democratic Republic of the Congo up-to-date on daily activities and situational awareness.[2] While not a documentary film project, this is an example of how geomedia has the potential to democratize marginalized communities and narrow the digital divide. It is also a good example of how an international body of changemakers such as the United Nations values such technology.

As we have examined in previous chapters, remediating the documentary form with the affordances of the digital domain has already taken place: the i-doc, the web-doc, the ecocinematic documentary, the multilinear documentary and the database documentary. Is it possible to unite these global structures and introduce them to a geomedia genre to modify the documentary progressively, to empower its ability to effect positive social change? Can a multilinear, interactive, database, social issue documentary film project intended to inform and influence power benefit from an exhibition platform of a GIS map? And if it can, will its activist intentions survive the process and contribute to a more effective and productive communications tool? In the following case study, we will examine how one experimental project, currently being used by the United Nations, addresses these questions.

[1] "About," *The Ushahidi Ecosystem*. www.ushahih.com. Accessed: August 17, 2018. https://www.ushahidi.com/about.

[2] "Performance Peacekeeping," *Keeping the Peace—The UN Department of Field Service's and Peacekeeping Operations use of Ushahidi*. www.ushahih.com. Accessed: August 17, 2018. https://www.ushahidi.com/blog/2018/08/08/keeping-the-peace-the-un-department-of-field-services-and-peacekeeping-operations-use-of-ushahidi.

7.2 Case Study: *The Youth Climate Report GIS Project*

By incorporating the research in the social issue documentary's ability to serve as an instrument of social change; its technological, epistemological, and artistic reforms since the dawn of moving images and cinema; and its transformations in all aspects of documentary production, exhibition and distribution in the digital age, I have created an experimental project to test the areas that have proven in theory and practice to best serve the activist intentions of the social issue documentary in an altogether new form I call the geo-doc.

The geo-doc is characterized primarily as a multilinear documentary film project presented on a platform of an interactive Geographic Information System map. The experimental project I built with this infrastructure is called the *Youth Climate Report GIS Project* (YCR) (Fig. 7.1). The YCR project, as its name suggests, is themed on the environmental issue of climate change, global in nature, and produced in an ecocinematic style, incorporating a mode engaging participation from the global communities of youth and science. Its GIS exhibition platform is a digital map of the world.

7.3 *YCR* Project Background

In 1972, the United Nations Conference on the Human Environment, more commonly known as the Stockholm Conference, established the United Nations Environment Programme (UNEP). Part of its organizing structure included a specific mandate to provide "an education programme designed to create the awareness which individuals should have of environmental issues" and that "(t)his programme will use traditional and contemporary mass media of communication" ("Recommendation 97," *Report of the United Nations Conference on the Human Environment*, 24).

Without expressly identifying film, and written in the days well before the common use of the Internet and other digital media, this mandate opened the doors for documentary film projects to participate in providing data related to environmental issues to policymakers and the public at large through UNEP. In her book, *Green Documentary: Environmental Documentary in the 21st Century*, Helen Hughes argues that this UN mandate was particularly influential in the rise of the eco-doc:

> For the environmental documentary the significance of this meeting (the Stockholm Conference) lies not only in the institutional developments but also in the rationalization and expansion of environmental assessment…It is this monitoring that has created a visible and legible data and insights that inform global discussions on the environment. It becomes particularly prominent in documentary films on climate change in the twenty-first century. (Hughes, 26)

This invitation to documentary filmmakers working with scientific themes is not unprecedented. Long before the UN existed, there was a movement to unite the world's scientists in the service of social justice by involving them in film. In 1935, an organization called the Associated Realist Film Producers was formed to establish contacts between scientists and documentary filmmakers (Boon, *Films of Fact*, 81–83). The organization had a panel of advisors comprised of filmmakers and scientists who were all active in a movement known as the "social relations of science" (Boon, *Science, Society and Documentary*, 323). Panel advisor and biologist Lancelot Hogben described his organization's goals as being essential in accelerating social progress:

> Money and effort spent in diffusing political propaganda might be far more usefully employed in promoting ad hoc societies to finance the production of documentary films dealing with specific social issues…The willing co-operation of men of science, the financial support of persons belonging to different political parties (or none at all) and the creative work of film directors…could be enlisted to quicken the social imagination. (Hogben, 6–9)

It was with these historical mandates that UNEP reached out to the documentary film community in 2009 to supplement written reports of polar science made during International Polar Year (March 2007 to March 2009).

The Youth Climate Report GIS Project would not have been possible without a previously established partnership with the United Nations. This section chronicles the history of that relationship that began with two eco-docs I made on the polar regions, *The Antarctica Challenge: A Global Warning* (2009) and *The Polar Explorer* (2010), and the role they played in bridging the gap between science and policy at the annual UN climate summits known as the COP conferences.

In 2008, International Polar Year (IPY) was concluding its two-year programme of intense scientific research at both ends of the earth. The programme, conducted over a two-year period from 2007 to 2009, was sponsored by the International Council for Science (ICSU—formerly known as the International Council of Scientific Unions) and the World Meteorological Organization. The chairs of the International Planning Group established within the ICSU for this event were Professor Chris Rapley, director of the British Antarctic Survey, and Dr Robin Bell, a geophysicist at Colombia University in New York.[3] In order to have full and equal coverage of both the Arctic and the Antarctic, IPY covered two full annual cycles from March 2007 to March 2009, and involved more than 200 projects, with thousands of scientists from more than sixty nations examining a wide range of physical, biological and social research topics.[4] In 2009, the United Nations Framework Convention on Climate Change (UNFCCC) was scheduled to take place in Copenhagen, Denmark, highlighting the extensive research done that year in the Arctic and Antarctica as part of its IPY commitment.

Eager to present as much polar data as possible at the COP15 conference, the UN reached out to the polar science community, many of whom were still embedded in their respective polar regions conducting research. Many of the scientists, particularly those in Antarctica, were not yet ready to release their findings and not available to attend, so they suggested that the conference organizers contact me, the filmmaker behind *The Antarctica Challenge: A Global Warning*, the only documentary film made in Antarctica during International Polar Year. The film and myself were invited to take on a new role: the feature-length documentary was to serve as a data delivery system to conference delegates, national delegations, NGOs, negotiators and policy writers, distinguishing it as "the only film invited by the United Nations to screen at the 2009 COP15 conference" (*Sioux Falls Scientists*, 2009), and I was to serve as a surrogate for the scientific community to facilitate communication between them and the environmental policymakers of the United Nations.

The reason for this privileged status for a commercially produced documentary is provided by the Under-Secretary General of the United Nations and Executive Director of the United Nations Environment Programme (UNEP), Achim Steiner:

[3] Wikipedia contributors. "International Polar Year." *Wikipedia, The Free Encyclopedia*. Wikipedia, The Free Encyclopedia, April 15, 2015. Web. Accessed June 26, 2015. https://en.wikipedia.org/w/index.php?title=Special:CiteThisPage&page=International_Polar_Year&id=656671702.

[4] "About IPY," *International Polar Year*. www.ipy.org. Web. Accessed June 26, 2015. http://www.ipy.org/index.php?/ipy/about/.

> Of all the canaries in the climate coal mine, the polar regions and the mountain glaciers are singing the hardest and the loudest. *The Antarctica Challenge: A Global Warning* underlines these realities with some of the latest and increasingly sobering scientific findings. (*Canadian Press/CTV News*, 2009)

The visual medium of film was well received by conference delegates that year, and as a result, conference organizers invited me to produce a second documentary film to screen in the same fashion at the follow-up conference, COP16, held in Cancun, Mexico, in 2010. The film that year was *The Polar Explorer*. It updated delegates on polar research in Antarctica and presented the first crossing of the Northwest Passage by polar scientists documented by film. Steiner once again praised the involvement of the documentary film in the policymaking process in a press release dated November 22, 2010, titled *UNEP Partners with Film Company for Climate Change Conference*:

> "The speed at which the polar regions are melting needs to be reflected in the speed with which nations come to agree on a decisive and definitive new global climate agreement. While the science is speaking loudly, it is often difficult for millions if not billions of people to witness this with their own eyes. This is the value of the film *The Polar Explorer*," Mr. Steiner added, "seen through the eyes of the scientists on the front-line, it brings the climate impacts at the poles to audiences in the conference halls of Cancun and the computers of the global public in order to raise the alarm but also the imperative to act." (UNEP, November 22, 2010)

The press release represents a significant relationship between the documentary filmmaker and global policymakers. Referring to the relationship as a partnership, the United Nations establishes a level of collaboration between the UN and a documentary film production company. As a partner programme, the project is used by delegates and negotiators in the policy creation process as a data delivery system. It is not simply an independent resource, but one created specifically with an audience of international policymakers in mind. The press release also identifies the participation of the global community of science in the project as well as the film's audiences: the policymakers in Cancun and the global public. Finally, and perhaps most importantly, the release calls for these audiences to act after seeing the film.

Consequently, action was indeed taken as a result of the screenings of *The Polar Explorer* at COP16. Following one of the screenings for the delegates of the European Union, I was invited into a policy-writing session on the subject of rising sea levels to review scenes of the film and be interviewed for further information as required. In this instance, the documentary filmmaker and the international policymaker worked together at the same table to draft environmental policy for the world. The film served as a resource of visible evidence for the two parties. We worked on a section of the *Cancun Accord* called "Enhanced Action on Adaptation." As a direct result of their screenings and meetings, they penned the following resolution known as Subsection 25 in the Accord's second section:

> (The Conference of Parties) (r)ecognizes the need to strengthen international cooperation and expertise in order to understand and reduce loss and damage associated with the adverse effects of climate change, including impacts related to extreme weather events and slow onset events, (i)ncluding sea level rise, increasing temperatures, ocean acidification, glacial retreat and related impacts. ("Decision," 6)

The events listed in the resolution were events documented in the film, showcasing recent research made by the scientists of the British Antarctic Survey and ArcticNet.

In recognizing the value of data visualization in traditional documentary format, the communications departments of UNEP and the UNFCCC requested annual film reports of climate-related research for each of their annual COP conferences going forward. Since the costs of documentary feature-length production are prohibitive for independent documentary filmmakers, the concept of the *Youth Climate Report* (see Fig. 7.2) was introduced as a viable alternative in 2011.

Fig. 7.2 Poster for *The Youth Climate Report*. Neko Harbour Entertainment, Inc., 2011–2015

7.4 CASE STUDY: *The Youth Climate Report* Film Series

The original structure of the YCR films included creating a new platform for the global communities of youth and science to have their voices heard at UN climate summits. Young people the world over were crowd-sourced by the communications department of the United Nations Environment Programme to produce short video reports of climate research in their respective regions. Scientific organizations were also recruited to showcase their latest work and to have their researchers interviewed by youth reporters. Six video reports, each running forty-five minutes in length, were produced between 2011 and 2015, and presented at the conferences of COP17 to COP21.

A remediation of this documentary format took place in 2015, introducing GIS as a presentation platform. This new approach in documentary filmmaking was welcomed by the delegates who reported to me that they found the system easy to use and informative. The previous chapters outline these methods, and a new version of the *Youth Climate Report* incorporating these ecocinematic, digital, multilinear, database, interactive, and geomedia components was created. Instead of one forty-five-minute, linear documentary feature showcasing approximately five reports, the database documentary composition now accommodated more than fifty reports in its first year. This new structure also afforded metadata combining the video reports with links to written texts of research, 360-degree photography, dates of the research, and locative coordinates of where the research was conducted. The global map[5] comprising these "mini-docs" also provided the opportunity for spatial and temporal analyses. Delegates attending the Paris climate summit in 2015 responded well to this augmented communications tool, praising its comprehensive data and easy interface. They requested more video reports going forward, and the content of the project grew from 50 film units to 181 in the following year. After addressing this request from delegates, the *Youth Climate Report GIS Project* was officially adopted as a data delivery system under Article 6 of the UNFCCC's mandate for education and outreach at the COP22 conference in Marrakech, Morocco, in 2016 ("Lights, Camera, Marrakech," 2016). Each year the map is updated on the UNFCCC

[5] YCR GIS Map on UN Climate Change website: https://unfccc.int/news/see-inspiring-climate-action-entries-to-global-youth-video-competition.

website. The latest version can be found on the web page for COP25, held in Madrid in 2019 ("Lights, Camera, Madrid," 2019).

7.5 *The Youth Climate Report GIS Project*: Methodology

7.5.1 *Components*

In remediating the structure of *The Youth Climate Report* (2011–2015) to a geo-doc, the basic concept of showcasing youth-as-reporters interviewing climate researchers-as-subjects was maintained for the *Youth Climate Report GIS Project* (see Fig. 7.1). The primary audience of UN climate conference delegates was also kept along with a secondary audience of the global public in accordance with Mr Steiner's requested mandate documented in his 2010 press release previously referenced.

The new elements consist of a presentation platform of a Geographic Information System map originally built with an open-source software called Google Fusion Tables. The platform existed in beta mode, and in December of 2018, Google announced it would discontinue this programme as of December 3, 2019.[6] To current users of this beta programme, Google has recommended migration to an existing platform, also open-source, called *My Maps*. Like Fusion Tables, *My Maps* offers a wide range of support for various digital media and layers for metadata. The metadata sets in the *Youth Climate Report GIS Project* (see Fig. 7.3) include the titles of the student-produced mini-docs, their filmmaker's names, the geographic coordinates of where the research is conducted, the

Fig. 7.3 Poster for *The Youth Climate Report GIS Project*, creator, Mark Terry, 2015–present

[6] Announcement: https://support.google.com/fusiontables/answer/9185417?hl=en.

year and country of production, and hypertext links to written texts of the research and/or links to the researcher and their respective science association: climate organizations, universities, NGOs.

These metadata do not need to be addressed during the viewing experience. With an average run-time of approximately three minutes per video, the entire "look and listen" spectatorship of the traditional documentary can be easily achieved with additional information available before or after the screening, much like "bonus material" is available on a commercially released DVD. As a result, they do not "interrupt" the screening process—a complaint of the Korsakow system—but rather augment it. The current incarnation of the project can be viewed and engaged with at this link: youthclimatereport.org.

7.5.2 *Curation*

In its 2015 beta mode, the *YCR GIS Project* curated its content through a combination of a social media campaign and direct recruitment from the communications departments of the UNFCCC, UNEP, and key science organizations (ArcticNet, British Antarctic Survey, David Suzuki Foundation, Global Foundation for Democracy and Development, the Royal Canadian Geographical Society, EcoCup Russia, The Explorers Club [sic], and the National Science Foundation), representing a comprehensive collaboration of changemakers and stakeholders.

After feedback from the communications departments of the UNFCCC and UNEP, this curatorial approach yielded fifty-four video reports from youth representing all seven continents. It was presented and tested by delegates attending the COP21 climate summit in Paris on November 15, 2015.

When the experimental project was taken out of beta mode, an augmented method of curation was introduced. The UNFCCC, in partnership with YCR and Television for the Environment (TVE), conducted a contest called the Global Youth Video Competition.[7] The contest invited the community of global youth aged eighteen to thirty to submit videos no longer than three minutes in duration, covering the climate issue of their choice. In subsequent years, the contest suggested two climate

[7] Global Youth Video Competition 2017. https://www.un.org/youthenvoy/2017/05/global-youth-video-competition-2017.

themes for filmmakers to follow, reflecting the key issues of focus at each respective conference. Within this framework, 181 videos were submitted in 2016, and each one was uploaded to the *YCR GIS Project*, together with its corresponding metadata.

This effective method of curation allowed for better recruitment from the communications departments of the UNFCCC and UNEP and TVE. The prizes awarded to the filmmakers were trips to attend the climate change conference and a certificate of achievement. The winning students and their films were introduced at two press conferences, one hosted by the UNFCCC and its partners, and one by the *YCR GIS Project*. This collaboration ushers in a new relationship between the changemaker audience and the documentary filmmaker on a global scale.

Since 2015, and as of September 2019, the project showcases 410 mini eco-docs on climate research from around the world. The subjects featured in these documentary projects are also active participants. Part of the curation process is to seek out qualified climate researchers and have them provide supportive footage of them working in the field or footage of the environment being impacted, and, in some cases, animated models of their research. Whenever possible, the subject engages the student reporter to film these segments to provide an enhanced understanding of the subject's work for the student. The "hands-on" experience provides a more robust background on the subject for the student reporter, resulting in a more informed interview for the policymaker.

7.5.3 *Production*

The actions of the documentary maker within the map project are relatively static and uniform. Student reporters are instructed simply to shoot an interview with a climate expert. An effort is made by the project's creator and his partners to curate videos from around the world and guide the student filmmaker so that the films have a similar appearance and structure. To ensure that all interested participants have the opportunity to produce, film, and contribute to the project, regardless of experience or knowledge in documentary production, training and guidance are available online at the project's website: http://youthclimatereport.org. Based on my own professional experience as a broadcast journalist and documentary filmmaker, I have devised a "tip sheet" provided on the project's website to assist the student filmmakers.

Interview Tips

1. At the beginning of your interview, ask the interview subject to state their name and spell it (our editor will need to know the correct spelling).
2. Ask them to state their title or profession and where they work (e.g. climatologist, University of Cambridge).
3. Ask them what new climate discoveries they have made that they feel delegates attending the climate change conference should know.
4. If you don't understand what they say, ask them to explain to you—remember, policymakers aren't necessarily scientists, and they will need explanations too.
5. Ask them what they believe is the greatest climate change crisis facing the world today.
6. Ask them if they have an idea how we can correct this problem or protect ourselves from its dangers.
7. Ask them to provide a message aimed directly at delegates and policymakers.[8]

Camera Tips

1. Position the camera on a tripod or similar stationary support (like a stack of books).
2. Frame the shot so the interview subject is seen from the top of his or her head down to the chest.
3. The interview subject should appear at one side of the frame (left or right). Be sure there is something interesting to look at in the empty space beside them (bookcase, computer, map, art, etc.).
4. Do not place your interview subject in front of a window.
5. Do not place your interview subject in front of a blank wall.
6. Place a remote microphone on the lapel or shirt of the interview subject for best sound results.
7. If you cannot get a remote microphone, get the camera and its mic as close as possible. Remember, preference will be given to those videos with the best production value.
8. Make sure the room and the surrounding area are as quiet as possible!
9. Email us the interview subject's name, title and affiliation or organization along with your video upload.[9]

[8] http://youthclimatereport.org/help/interview-tips/.
[9] http://youthclimatereport.org/help/camera-tips/.

7.5.4 *Collaboration*

With collaborating organizations such as the UN Climate Change secretariat, UNEP, UNDP, and TVE as producers; international universities and science organizations, as interview subjects and content providers; and the global community of youth, as reporters and filmmakers, international collaboration on this project allows for a comprehensive curation of climate science, interview subjects, and student reporters. It is important to note here that all curated videos remain on the map permanently while new ones are added continuously. This provides an historical perspective of the research, so users of the project can make temporal as well as spatial analyses of the data. Assuming the same location and research subject, this is important in assisting researchers and policymakers in making future projections.

7.5.5 *Technical Systems*

The technical systems involved are primarily those used in any documentary film project: a camera (digital), audio recording devices (microphones, lavaliers), lighting (both natural and artificial) and in addition to these are the tools available within the digital domain: editing software, file transfer sites, hard drives and archive sites such as Google Drive, UN websites and YCR websites. For exhibition, in addition to the distribution of links to view the multilinear project, a large touch-screen monitor is set up at the COP conferences so all delegates passing by may engage with it and bring the information they receive to their respective policy meetings. Delegates can also access the map online using their laptops during negotiating and writing meetings.

7.5.6 *Social Media*

The social media established to serve the *Youth Climate Report* in its original linear documentary format saw modest increases in subscriptions following a tour hosted by participating nations to recruit and train youth reporters in 2011. These numbers were significantly increased when the project evolved into its multilinear, GIS format. Here are the numbers in the respective eras of each format:

- Twitter—https://twitter.com/ycrtv
  YCR Linear (2011–2015): 112; *YCR Multilinear* (2015–2020): 855 Followers
- Facebook—https://www.facebook.com/youthclimatereport

YCR Linear (2011–2015): 368; *YCR Multilinear* (2015–2020): 2896 Followers

- YouTube—https://www.youtube.com/youthclimatereport

YCR Linear (2011–2015): 18 Subscribers; *YCR Multilinear* (2015–2020): 379 Subscribers.

It is likely that the reason for the increase in subscriptions relates not only to a format that involves more participants, but also to the fact that the communications departments of the United Nations coordinate an annual call for entries and award the winning submissions with prizes of travel and participation in the COP conferences.

7.6 *The Youth Climate Report GIS Project*: Applied Theory

7.6.1 Introduction

With the social issue documentary remediated in this manner, the various theories examined in previous chapters that demonstrate successful practice in influencing the changemaker and effecting social change are applied in this experimental geo-doc project. Specifically, the *Youth Climate Report GIS Project* applies the following theories:

- Communications tool
- Community participation
- Collaboration with the changemaker
- Presenting directly to the changemaker
- Ecocinema
- Digital affordances: Geomedia, GIS platform, narrowing the digital divide, and spatial and temporal analyses
- Multilinear format

7.6.2 Communications Tool

In Chap. 2 we examined the early days of the documentary film and found how the medium of film, in its non-fiction format, exhibited an innate ability to inform, educate, and influence. In 1897, two years after the technology of moving pictures was introduced to the world by Lumière brothers, James Freer, a farmer, played a significant role in stimulating immigration to Canada. Recognizing the communicative power of this new technology, Clifford Sifton, Canada's Minister of the Interior at the time, recruited Freer, and later Charles Urban, to continue to create promotional films showcasing Canada as an attractive destination for immigrants. In Europe, Urban developed his own catalogue of non-fiction films on the subject of science as well as military training, both in the service of educators. In the 1920s and 1930s, documentary pioneers Joris Ivens, Dziga Vertov, and John Grierson used film to amplify social injustices, not only to inform the general public, but to persuade changemakers to improve the conditions of those profiled in these early social issue documentaries.

This inherent ability to speak truth to power defines the documentary film as a valuable communications tool. For this reason, the *Youth Climate Report GIS Project* identifies the communications departments of UNEP, UNDP, UNESCO, and the UN Climate Change secretariat as both collaborators and audience on a subject of science to assist the international government body in creating global environmental policy. Remediating the documentary as a digital communications tool for the communications department of UN agencies allows this geo-doc to be designed specific to the needs of the changemaker, ensuring its effectiveness as a digital instrument of social change.

7.6.3 Community Participation

As early as 1933, we saw filmmakers like Ivens recruiting community participation in his film *Misère au Borinage* so that the profiled miners suffering inhumane working conditions could best represent their struggle to the public as well as their employers. Thirty years later, we saw the same approach taken by Father Albert Tessier, handing over his cameras to the rural communities of northern Quebec so they may tell their own stories of economic hardship rather than have the filmmaker interpret their struggles. This prompted Colin Low to introduce the same technique with the impoverished fishing community of Fogo Island. He presented films comprised of footage shot by the local community directly to the federal government resulting in the establishing of a co-op designed specifically to help the people of Fogo Island. In 2003, George Ferreira used digital technology to have the Keewaytinook-Okimakanak communities in Northwestern Ontario record and stream their stories of economic hardship directly to the federal agency of Industry Canada to assist the policymakers there to institute financial relief programmes.

This technique in the Canadian tradition demonstrates the documentary film's power to persuade. The success of this participatory mode of documentary filmmaking inspired a similar approach to the production of the *Youth Climate Report GIS Project*. The global community of youth is asked to record researchers from the global community of science to have the complicated data of climate change explained in a more accessible way, a way that the less scientifically knowledgeable policymakers of the United Nations find easier to understand. Additionally, the profiled science community is asked to showcase its findings and research methodologies to student reporters ensuring that the most significant data are highlighted, according to the world's foremost experts in the field. And finally, the global policymaking community of the United Nations can suggest presentation formats, research themes and other content they deem necessary to assist them in specific policy creation meetings. The result is a collaborative effort involving the participation of the global communities of youth as filmmakers, science as interview subjects, and policy as changemaker.

7.6.4 *Collaboration with the Changemaker*

One of the key components we examined in the success of any interactive, digital documentary is collaboration. Working together with profiled communities assists the filmmaker in representing the profiled social issue from the perspective of those most impacted. The relationship between the filmmaker and the changemaker in this regard yields promising results in enhancing the documentary's ability to serve as an instrument of social change. Having audiences collaborate on content creation, online commenting, and digital dissemination such as sharing, magnifies the message and extends its reach. As the authors of *The Act of Documenting: Documentary Film in the 21st Century* point out in their chapter entitled "Giving Voice":

> If social engagement is to be a factor, then collaboration and participation in filmmaking, it is assumed here, must mean more than mere involvement. In this context, the degree of potential redistribution of the filmmaker's traditional agenda setting power is the measure of both. (Brian Winston et al., 112)

In this regard, one of the strongest features of the *Youth Climate Report GIS Project* is its collaborative nature with its specific audience, the change-maker. Key stakeholders in the communications department of UNEP and the United Nations Climate Change secretariat were instrumental in designing the architecture of the project. Their specific requests for meta-data, climate-themed data, and a multilinear assembly of video reports on a single, global platform ensured that the changemaker was provided with everything they identified as being required for them to best craft policy based on as much relevant data as possible. The semiotic nature of GIS provides an extra dimension of comprehension to the changemaker through its unique affordances of spatial and temporal analyses. Since these data were requested by the changemaker, it can follow that a fuller understanding of the climate research is more available than in traditional documentaries that do not provide it.

7.6.5 Presenting Directly to the Changemaker

The primary audience of the *Youth Climate Report GIS Project* are the changemakers, specifically the policymakers of the United Nations, not the general public. The public at large is usually the target audience of the makers of traditional social issue documentaries. These documentarians aim to engage emotionally these public audiences with the goal of motivating *them* to influence the changemaker to create social change. This method has some measure of success—such as the examples we saw in the films *Blackfish* and *Sin by Silence*—however, we only know they were effective because the changemaker directly attributed the influence of these films to their decisions.

As we saw with the films of Low and Ferreira, as well as the examples of *The Price We Pay* and *The Polar Explorer*, when policy is changed because of a documentary film, it is always attributed to the films that were screened *directly* to the changemaker. For this reason, I arrange for the *Youth Climate Report GIS Project* to be presented each year at the UN climate summits directly to delegates, negotiators, and policy writers. The project is used by the UN as a data delivery system to assist them in creating fully informed environmental policy for the global communities they serve. It is hoped that future geo-doc filmmakers follow this important aspect of project presentation to maximize its power to inform and influence those directly charged with creating positive social change.

7.6.6 Ecocinema

In Chap. 4, we examined how a particular sub-genre of the documentary, the eco-doc, with its global intentions to inform and influence, is imbued with semiotic storytelling affordances and techniques such as the anticipatory mode, and implicit narratives evident in production approaches designed to retrain our perception about what we see on the screen. This sub-genre and its methods provide additional data and influential power necessary to accelerate the action required to be taken by the global audiences of environmental policymakers as well as the global public to change the way we engage with our planet. The goal is to, in fact, retrain our perception not just of film, but also of our relationship with the environment, locally and globally.

For these reasons, the *Youth Climate Report GIS Project* is comprised of eco-docs that employ the semiotic techniques of ecocinema, not necessarily in its approaches to the production of the individual film units, but in the overall approach to production of the multilinear, GIS film project through the spatial and temporal analyses it affords its users. As well, these eco-doc film units represent a myriad of environmental issues related to climate change and presented on a global scale. This representation of content mirrors the all-inclusive nature and global view taken by the international policymakers of the United Nations whom this project targets as its primary audience. It is with their collaboration, and the participation of the global communities of youth and science, that the *Youth Climate Report GIS Project* displays its greatest strength as an ecocinematic communications tool, one so robust, in fact, that it has been adopted by the UN as a partner programme under its Article 6 mandate for education and outreach.

7.6.7 Digital Affordances

As we saw in Chaps. 5 and 6, and earlier in this chapter as well, the documentary film has been remediated several times and in many ways, taking advantage of the unique affordances available in the digital domain that the documentary did not have in its previous and relatively restrictive state of celluloid and cinema. Today's digital affordances now provide wider distribution of content through digital dissemination and social media sharing; an enhanced mode of participation through the digital collaboration of audiences and changemakers before, during and after production;

and the ability to reach millions of audience members simultaneously, not just the few hundred confined to the physical structure of a cinema. As well, digital geomedia provides a platform that merges electronic media with locative media to augment the documentary's storytelling and informative powers enabling implicit narratives that yield greater context and subsequent comprehension of the profiled social issue. The digital divide is also narrowed by curating global communities of science and youth and their most accessible camera technologies with step-by-step instructions to be as inclusive as possible.

I have incorporated all of these digital affordances, and more, into the *Youth Climate Report GIS Project* geo-doc. By placing hundreds of video reports of environmental issues worldwide directly into the hands of UN policymakers in a geomedia project of this nature, additional digital affordances—such as access to metadata and spatial and temporal analyses—all available in one digital place, remediate the documentary in a way that is most informative to the changemaker, and in a manner the changemaker himself has identified as being most helpful. The geo-doc, therefore, can be seen as being more influential to the changemaker than traditional documentary approaches to storytelling since they helped create it and design specific features and content they require. The geo-doc's extended use of digital affordances available in a geomedia platform also augments the traditional documentary film's innate ability to inform and influence.

7.6.8 Multilinear Format

In Chap. 6, we examined another digital remediation of the documentary: the multilinear format. The multilinear documentary film project presents several film fragments—or Smallest Narrative Units (SNUs)—in a single, digital space, usually a website. From here, audiences act as their own editor selecting these SNUs in an order they determine. While this format provides a wide variety of content to view and gives the audience the power to choose which film fragment to watch, we have examined a significant amount of criticism concerning the format's ability to maintain a narrative and to establish a comprehensible context of its content to its audiences.

For these reasons, I deliberately avoided the reliant nature of the film fragments shot together—but made available separately—to address this contextual concern. In this geo-doc project, the database of all film units is stand-alone mini-docs, not puzzle pieces that run the risk of never being assembled coherently by the project's user. The inter-relationality of the film units in a multilinear format still exists, but within a geomedia platform, they serve to provide additional contexts, those provided by spatial and temporal analyses. The project is also presented not on a "flat" website, but a GIS map of the world, providing a constant reminder to the user/audience of the global nature of the presented content. In this particular multilinear format, the entire experience of just "looking at and listening to" a traditional documentary is maintained, while at the same time making available additional context, data, photographs, and perspective not available in linear documentaries or traditional multilinear documentary film projects.

7.7 Conclusion

This chapter introduced a geomedia project that remediates the documentary in a specific way, a way that incorporates the collaboration of the changemaker to provide a communications tool that addresses their need for specific content. If the documentary film has been used in the past as an effective instrument of social change, this comprehensive, digital, and collaborative approach should strengthen its ability to serve in this capacity.

But does it work? In its infancy in 2015, the *Youth Climate Report GIS Project* was presented on a wall of touch-screen monitors with headsets enabling all delegates of the Paris climate summit, COP21, to engage with it. NGOs and youth groups gravitated to the publicly interactive nature of the new medium and introduced it at side events as a new curative communications tool for their stakeholders to use and to contribute to. In addition to the public installation of open-access monitors, negotiators and policy writers were given access to the project on a secure server so they can work with it using their computers, privately or in meetings. It was from this use in particular that the project received its greatest input from its targeted audience of changemakers.

Nick Nuttall, head of Communications for the United Nations Climate Change secretariat, issued a press release praising the conference's new communications tool:

> The many videos we received from around the world, placed … on this global, easy to use map, are testimony to (youth's) commitment, and I'd

> encourage everyone to take a look at (the) numerous examples of climate action on the ground (that) will be showcased at the upcoming UN Climate Change Conference in Marrakech in November, and the videos help underscore the fact that governments, cities, regions, businesses and investors are all stepping up to plate to jointly tackle climate change and create a greener, safer and more sustainable future for themselves and the world. ("See Inspiring Climate Action," 2016)

When the press release was issued, it was followed by a press conference one month later at the next climate summit, COP22, in Marrakech, Morocco. The 2016 version of the *Youth Climate Report GIS Project* incorporated the changes requested by the UN and was populated by 181 new video reports curated by the *Global Youth Video Competition*. It was at this press conference that it was announced that the experimental geo-doc project was now an official partner programme of the UN Climate Change secretariat adopted under its Article 6 mandate for education and outreach (*Lights, Camera, Marrakech*, 2016).

Since then, other UN partner agencies, such as UNEP, have engaged with the project and commented on its value as an effective communications tool. In a 2018 interview, Executive Director of UNEP, Erik Solheim, praised the project for bridging the gap between science and policy:

> It comes down to a need to find the best methods for communicating science. The most important thing to remember is that the decision-makers are not scientists. Certainly, they have a grasp of the basic concepts, but politicians and civil servants are also generalists. Data visualisation in (the *Youth Climate Report GIS Project*) is one of the incredible tools available to get the science and the data across. ("Interview with Erik Solheim," 2018)

In addressing the value of giving a voice to the global community of youth, Solheim says that the *Youth Climate Report GIS Project* recognizes that "(e)very voice counts and every voice must be heard. Climate action is about leaving behind a safe planet for future generations. Every delegate needs to be reminded of that moral obligation" ("Interview with Erik Solheim," 2018).

Ecocinema documentary film theorist Helen Hughes also argues that the *Youth Climate Report GIS Project* is a valuable communications tool for both scientists and policymakers:

> (The project) has created an opportunity for young people and scientists all over the world to express their activities with respect to caring for the environment. (The project) has opened a window on the many ways in which people have been thinking about the issues. Looking at the videos from across the world there is a wonderful diversity of approach both to video making and to the kinds of science relevant to the problem of the environment and climate change research. (The films represent) models of clear, accurate, visually arresting communication… these films (show) how important such vehicles can be for the policy making process. (T)he UN online project is the most amazing visionary experiment. ("Analysis of the United Nation's Youth Climate Report GIS Project," 2017)

In addition to the direct connection made between Resolution 25 of the Enhanced Action on Adaptation section of the Cancun Accord and the documentary feature *The Polar Explorer* (2010), this remediated version of the social issue documentary, made in collaboration with the international environmental policymakers of the United Nations in 2015 and adopted as an official data delivery system in 2016, continues to provide evidence of the documentary's direct influence on the UN's policy creation process in the arena of climate science.

In terms of impact theory and engagement, the geo-doc has demonstrated a significant potential to influence by providing more data—both implicit and explicit—in a unique way. As users of the experimental project have indicated, they stress the communicative powers of the geo-doc as being most valuable. By collaborating with the changemaker before, during, and after the filmmaking process, required information is assured, and through the affordances provided by digital and geomedia means, that information is presented in a manner that provides a fuller understanding than a traditional documentary film with a limited run-time.

By compiling the proven successful methods, theories, techniques, practices, and technologies that the social issue documentary has employed over the years into a new mediated version as we have examined in this case study, we can use the geo-doc as a model for future documentary projects. These projects can be local or regional and still provide spatial and temporal analysis, metadata such as photography and written reports, visible evidence in the form of documentary footage, and live camera feeds, all necessary components that imbue the geo-doc with the potential to deliver a fuller understanding of a social issue. On a global scale, the structure and the accessibility of the geo-doc provide a new way for the film genre of the documentary to inform and influence the changemaker who is charged with creating policy that results in positive social change on issues that impact all of us.

Bibliography

"About." *The Ushahidi Ecosystem*. www.ushahih.com. Web. Accessed August 17, 2018. https://www.ushahidi.com/about.

"Analysis of the United Nation's Youth Climate Report GIS Project." *youthclimatereport.org*, September, 2017. Web. Accessed May 19, 2018. http://youthclimatereport.org/news-cat/analysis-united-nations-youth-climate-report-gis-project-helen-hughes.

"Canadian Film to Be Shown During UN Climate Meeting." *Canadian Press / CTV News*, August 18, 2009. Web. Accessed May 13, 2018. https://www.ctvnews.ca/canadian-film-to-be-shown-during-un-climate-meeting-1.426585.

"Decision Adopted by the Conference of Parties." *Cancun: United Nations Framework Convention on Climate Change*, March 15, 2011. Web. Accessed May 5, 2018. http://unfccc.int/resource/docs/2010/cop16/eng/07a01.pdf#page=4.

"Interview with Erik Solheim." *youthclimatereport.org*, April, 2018. Web. Accessed May 19, 2018. http://youthclimatereport.org/?p=2428&preview=true.

"Lights, Camera, Madrid." *United Nations Climate Change. Madrid, Spain*, Novemvber 9, 2019. Web. Accessed January 12, 2020. https://unfccc.int/topics/education-youth/youth-engagement/global-youth-video-competition/lights-camera-madrid.

"Lights, Camera, Marrakech." *United Nations Climate Change*. Marrakech, Morocco, November 9, 2016. Web. Accessed May 14, 2017. https://unfccc.int/topics/education-and-outreach/events%2D%2Dmeetings/global-youth-video-competition/2016-global-youth-video-competition-on-climate-change.

"Performance Peacekeeping." *Keeping the Peace – The UN Department of Field Service's and Peacekeeping Operations Use of Ushahidi*. www.ushahih.com. Web. Accessed August 17, 2018. https://www.ushahidi.com/blog/2018/08/08/keeping-the-peace-the-un-department-of-field-services-and-peacekeeping-operations-use-of-ushahidi.

"Recommendation 97." *Report of the United Nations Conference on the Human Environment*. Stockholm: United Nations, 1972. Print.

"See Inspiring Climate Action." *Resource: United Nations Framework Convention on Climate Change*, October 19, 2016. Web. Accessed May 19, 2018. https://unfccc.int/news/see-inspiring-climate-action-entries-to-global-youth-video-competition.

"The Antarctica Challenge: A Global Warning." *Sioux Falls Scientists*, n.d. Web. Accessed May 13, 2018. http://www.siouxfallsscientists.com/the-antarctica-challenge.html.

Boon, Tim. *Films of Fact: A History of Science in Documentary Films and Television*. London: Wallflower, 2007. Print.

———. "Science, Society and Documentary." In *The Documentary Film Book*, ed. Brian Winston. London: British Film Institute, 2013. Print.

Hallam, Julia, and Les Roberts, eds. *Locating the Moving Image: New Approaches to Film and Place*. Bloomington, IN: Indiana University Press, 2014. Print.

Hogben, Lancelot. "The New Visual Culture." *Sight & Sound* 5 (Spring, 1936). Print.

Hughes, Helen. *Green Documentary: Environmental Documentary in the 21st Century*. Bristol, UK: Intellect, 2014. Print.

Lapenta, Francesco. "Geomedia: On Location-Based Media, the Changing Status of Collective Image Production and the Emergence of Social Navigation Systems." *Visual Studies* 26, no. 1 (2011): 14–24. Print.

McQuire, Scott. *Geomedia: Networked Cities and the Future of Public Space*. Cambridge: Polity Press, 2016. Print.

Miles, Adrian. "Programmatic Statements for a Facetted Videography." In *Video Vortex Reader: Responses to YouTube*, ed. Geert Lovink and Sabine Niederer. Amsterdam: Institute of Network Cultures and XS4All, 2008. Print.

Thielmann, Thomas. "'You Have Reached Your Destination!' Position, Positioning and Superpositioning of Space through Car Navigation Systems." *Social Geography* 2, no. 1 (2007): 63–75. Print.

UNEP Partners with Film Company for Climate Change Conference. United Nations Environment Programme Press Release. Cancun: United Nations Environment Programme, November 22, 2010. Web. Accessed September 7, 2016.

Winston, Brian, Gail Vanstone, and Wang Chi. "Giving Voice." In *The Act of Documenting: Documentary Film in the 21st Century*. New York: Bloomsbury Publishing Inc., 2017. Print.

CHAPTER 8

Conclusion

Fig. 8.1 *Filmmaker Mark Terry at COP21*. Hollywood Canada Communications, 2015

Filmmaker as Policymaker? The lines are blurring between those who produce social issue documentaries with a goal of influencing policy and those who are influenced by documentary film to create policy. Documentarians are now being given a seat at the table to assist in the creation of policy through the collaborative and co-productionary contributions of the changemaker. Seen here: Documentarian Mark Terry as a Party Delegate representing Canada at the Paris Climate Conference, COP21 in 2015.

M. Terry, *The Geo-Doc*, Palgrave Studies in Media and Environmental Communication,
https://doi.org/10.1007/978-3-030-32508-4_8

This book set out to investigate the nature of the documentary in terms of its relationship with social change. In this book's exploration of how it has the ability to influence, we investigated its claims to the truth and found advances created by new technologies that help enhance the documentary's presentation of the truth through the affordances provided by digital technologies and GIS mapping as well as new presentation approaches provided by ecocinema, multilinear formats, audience collaboration, and direct engagement with the changemaker. These key approaches and digital technologies are combined to remediate the documentary into a form that promises to maximize its potential to effect social change: the Geo-Doc.

The *Youth Climate Report GIS Project* represents an experiment in combining the multilinear documentary with geomedia. By incorporating a GIS platform, the changemaker is better served to access the scientific research they require to inform and assist them in creating international environmental policy. However, this initial attempt to remediate the documentary in collaboration with the changemaker is far from complete. There exist more digital affordances that would be helpful to include. As an enhancement to providing digital photography of the profiled area of research currently available for each mini-doc, a live feed of the area would provide the ultimate temporal view of how the researched environment looks today. As a live video feed, its surveillant nature provides a real-time opportunity for the policymaker to encounter the area. The extended opportunity for two-way discussions between policymaker and the scientist in the field exists with the embedding of telecommunications applications such as Skype and Zoom.

This live option might also make possible scheduled interviews with climate researchers to provide further clarity to the UN climate change conference delegates should they have questions or require additional information not available within the film units and their related metadata of the project. This renovation to the digital architecture of the geo-doc would require cameras on the other end to be set up and active for the "real-time, any-time" option to work and be of value to the changemaker. Some national parks in the US, for example, have set up live video cameras permanently positioned on eagles' nests.[1] This feature, currently available as separate applications, widgets, and add-ons, can be added to most currently existing GIS software. The Stream widget, for example, is a live video feed app designed for ArcGIS projects.[2] Using this technology requires a collaboration between the scientific community and the policymaker to identify what live feeds are required and how best to establish them so they are reliable.

It is important to note that the geo-doc is not meant to replace existing communications materials such as written reports from the scientific community, but rather supplement these materials with additional media comprehensively contained in one easily accessible and navigable project. As well, the geo-doc is not structured in such a way that it only serves the issue of climate change and the environmental policymakers of the United Nations. In its current design, a curation of mini-docs on other global issues such as domestic abuse, hunger, poverty, transportation, to name just a few, can take advantage of the format in much the same way.

8.1 Virtual Reality

Further extensions of digital affordances, outside of yet inspired by the geo-doc, are also available and desired by changemakers like the United Nations. This global audience of policymakers have inquired about new media such as virtual reality (VR) to augment data visualizations and comprehension of scientifically complex information. While still in its infancy, VR projects have been produced for the UN and are available on their United Nations Virtual Reality (UNVR) website.[3]

The UNVR was established in January of 2015, coordinated by the United Nations Sustainable Goals Action Campaign, "to bring the world's most pressing challenges home to decision makers and global citizens around the world, pushing the bounds of empathy" (About UNVR, 2015). Echoing the intentions and goals of the *Youth Climate Report GIS Project*, the UNVR showcases non-fiction stories of those suffering from social, political, economic, and environmental challenges:

[1] Big Bear Lake Eagle Cam: https://www.youtube.com/watch?v=HhTd-_rj-d0; The Decorah Eagles: https://www.youtube.com/watch?v=R4b_1rUNCUY; Channel Islands National Park Eagle Cam: https://www.youtube.com/watch?v=35ppQYptEfU.

[2] Stream for ArcGIS: https://doc.arcgis.com/en/web-appbuilder/create-apps/widget-stream.htm.

[3] http://unvr.sdgactioncampaign.org/.

> Building upon its mandate to amplify the voices of those who are often unheard, particularly the world's most vulnerable, the project seeks to show the human story behind development challenges, allowing people with the power to make a difference have a deeper understanding of their world, and hopefully to act to make a difference. ("About UNVR," 2015)

By "amplifying the voices of those who are often unheard" and by directly showcasing their stories to those in power "at high-level UN meetings," the influence of the documentary film in both projects reveals its innate ability to effect social change through community participation, change-maker collaboration, and presenting the documentary media directly to those with the power to create social change, all key approaches I have argued as being essential to mobilizing the documentary film as an effective instrument of social change.

To date, UNVR makes available a total of twenty VR films ranging in duration from two minutes and forty-one seconds—*Building Brains, Building Futures* (2017)—to ten minutes and twenty-eight seconds: *Life in the Time of Refuge* (2016); so, while the empathetic engagement might be unique to the VR documentary, it is hindered by restrictions on content caused by the technology's preferred time limits. It appears that brevity is the soul of VR. In a 2016 study conducted by Fernando Tarnogol, the average runtime among forty 360 VR projects surveyed was only one minute and seven seconds. Projects longer than two minutes began to show gradual disengagement with their audiences, suggesting the medium may be distracting from the message. Content appears not to be looked at or listened to as much as the immersive digital environment is, and after everything is seen and heard within this 360-degree construct, the user feels that the story is over:

> (B)ased on the <2 minute analytics, as long as we stayed close to the 1 minute mark we are safe. One theory I'm pondering is that if the viewer is past the 1 minute mark and believes that the video is about to end, they will give you those extra 10 to 20 seconds to close the story up. Meaning that if we are shooting for a 1 minute total duration but due to the content's characteristics you can't avoid getting past the mark, generate some tension/anticipation on the mark to buy a few more seconds of attention. (Tarnogol 2016)

The 360-degree video format of VR has the advantage of not requiring special gridded rooms, sensors, headsets, and handsets, like the fully immersive environments created from scratch incorporating many media such as animation, video, computer-generated imagery (CGI) effects, photography, and even written texts. These VR experiences allow the user to traverse the terrain by walking and even jumping to trigger action within the digital environment. The 360-degree video is much more accessible through mobile devices, tablets and laptops, but the engagement time seems to be quite short. Tarnogol includes in his report the average amount of time users that he tested stay engaged with the 360-degree video VR experiences based on a variety of devices:

- Mobile: 56% (Average View Duration: 0:48)
- Computer: 39% (Average View Duration: 0:49)
- Tablet: 4.6% (Average View Duration: 0:49)
- TV: 0.2% (Average View Duration: 0:20)
- Game Console: 0.1% (Average View Duration: 0:17)

Almost all users chose either the mobile device or the computer as their delivery system, and perhaps most telling is that their average duration of engagement is less than fifty seconds for a two-minute VR film, less than half of its runtime. This is the equivalent of walking out of a cinema forty-five minutes into a ninety-minute feature documentary film.

Why are these kinds of VR experiences so under-engaged? The UNVR projects are comprised solely of 360-degree video of selected locations around the world telling non-fiction stories of the people who live there. They are also accessible on websites like YouTube, and do not require headsets and handsets that many other more immersive VR experiences require. In *Clouds over Sidra*, the user can visit virtually a Syrian refugee camp in Jordan and experience the living conditions of people escaping their war-torn home country.[4] The idea of this kind of data visualization is to provide empathy, an extended mode of audience engagement for the social issue documentary. But how successful is this new form of audience interaction, especially when the average view time is less than 50% of the film's already short runtime?

In a 2016 interview with new media theorist Henry Jenkins, comparative media studies scholar William Uricchio believes VR is a step in the right direction, but the technology still needs to be developed:

[4] *Clouds over Sidra* (2015). http://unvr.sdgactioncampaign.org//virtual-reality/cloudsoversidra/.

> VR can be a great attention-getter, a quick and easy way to create a sense of presence and place. By creating the impression of being somewhere, by giving the viewer the freedom to look up, down and all around, a lot of crucial contextual information can be derived that would, in more limited linear scenarios, require careful selection and plotting, only to wind up giving us the director's or writer's point of view. Immersion can offer a counterweight to indifference. It can lure us into being interested in a topic we might otherwise gloss over, can encourage a search for facts, or a desire to learn. Rational debate, as a mode of discourse, is usually driven by some sort of motive. Immersion can help to create that motive, but—at least until we develop better ways of shaping and directing immersive experiences—it is not, in itself, a mode of discourse. (Jenkins 2016)

While empathy is a valuable goal to achieve in social issue documentary film engagement, the current technology—in its most accessible form—is insufficient in providing the other proven characteristics (participation, semiotics, metadata) defined in previous chapters and demonstrated through the *Youth Climate Report GIS Project* as being equally valuable in informing and influencing the changemaker.

8.2 Antarctica in Decline

To that end, I have created a new VR project, incorporating the fully immersive experience of a manufactured landscape with many of the tenets of successful documentary filmmaking, as an experiment in remediating the documentary in yet another way with the affordances of digital technology. *Antarctica in Decline* (see Fig. 8.2) is a fully immersive VR project that brings users to the Larsen C ice shelf on Antarctica's northwest peninsula using an Oculus Rift system and built with Vive software powered by the Unreal Engine, a publicly available freeware used to create fully immersive VR experiences.

This VR project, designed in collaboration with the United Nations Climate Change secretariat, presents a recreation of the collapse of the Larsen C ice shelf which occurred on July 12, 2017.[5] It presents the user with an environment in Antarctica that they can explore, complete with snow, ice, wind, mountains, glaciers, and penguins. To provide some informational context, I created a shack (see Fig. 8.3) comprised of wood salvaged from the wreck of the Endurance, Ernest Shackleton's ship

[5] https://phys.org/news/2017-08-larsen-c-iceberg-breakaway.html.

Fig. 8.2 Poster for *Antarctica in Decline*. Designed by Mark Terry, 2017

featured in the 1919 documentary *South*. In the shack is a walkie talkie where users can hear the actual voice of Shackleton reporting on findings he made in the area more than a century ago and recreated in this VR experience. Also available in the room are maps, pictures, and information to inform the user of the geographic, historical, and climate-related data of the area.

Outside the shack is a floating information balloon. For this, I created a proximity trigger. When the user gets close enough to it, they activate a video of modern-day Antarctic scientists who report on the Larsen C ice shelf. Walking further on, the user encounters the crack described by the scientists and which is visible in the pictures and maps from the shack. If

Fig. 8.3 Scene from *Antarctica in Decline.* Designed by Mark Terry, 2017. https://youtu.be/Siaqtkxatv8. Frame of scene occurs at 00:03:17

the user approaches too closely, they will trigger the collapse of the ice shelf in dramatic fashion.

The purpose of this project is to provide the user—in this case, the environmental policymakers of the United Nations—with an empathetic experience to accompany the data they have received in written and film texts. I chose Antarctica as a destination since very few people have visited this remote continent. As well, Antarctica's unique climate and terrain are unfamiliar to most in terms of first-hand encounters. This digitally enhanced experience aims at providing the policymaker with a more complete understanding of climate issues affecting this continent and, by extension, the world. It also remediates the documentary by including the familiar media of video, audio, and text within the relatively new media of immersive 360-degree virtual environments. Unlike the 360-degree video films we examined previously, the extended immersive environment of a world created entirely with digital technology has no set duration. The user can traverse the frozen terrain and experience the embedded multimedia for as long as they like, allowing for further study, review, and exploration. As an instrument of social change, this remediated version of the documentary provides the same amount of data and metadata found in a geo-doc, but adds an extra dimension of audience engagement: empathy. This unique affordance provided by VR heightens the audience's engagement with and understanding of the social issue presented.

Not all new documentary technology is as sophisticated as VR. Current filmmaking tools have been democratized in terms of accessibility and affordability. The mobile device many of us carry in our pockets has high-quality video-recording capabilities, and mobile applications allow for editing to refine those images and compile them in traditional narrative formats. This results in an expansion of participation in the documentary filmmaking world, sometimes competing right alongside the professional documentarian for that coveted slot at a film festival or for licensing by a broadcaster or video streaming service. The abundance of content this produces includes more unprofessional projects than we have seen before as the amateur filmmaker can fail to produce "broadcast quality" footage in the absence of formal training, even when their story might be of broadcast significance. This does not mean, however, that these projects are to be summarily dismissed. Even poorly produced documentary fragments on their own can contribute directly to progressive social changes. Police officers found guilty of brutality, for example, have lost their jobs after footage is released of their beatings of suspects. These videos are often taken by amateur filmmakers, bystanders to the incident, and provide visible evidence to a crime and sometimes to a greater, more systemic problem. In a 2015 paper written by Ben Brucato titled *Policing Made Visible: Mobile Technologies and the Importance of Point of View*, it is argued that this ubiquitous "new transparency" not only makes criminals easier to catch and convict, but it places an accountability on potential criminals in positions of power (law enforcement, political office, and corporations) that could reduce future occurrences of the crime, resulting in a progressive reduction in crime for society among those not usually accountable:

> Those living in surveillance societies have inherited a modern, Enlightenment doctrine of transparency that has been modified by broadly diffused imaging technologies and widespread participatory media practices….Video documentation of official conduct—and misconduct—has promoted broader visibility of political and corporate institutions and the agents therein, and this

visibility is now identified with the modern concept of transparency. (Brucato, 457–458)

A geo-doc platform of videos of this nature can provide the visible evidence required to tackle widespread issues among hegemonic institutions previously seen to be immune from investigation and prosecution. The national or global perspective offered by a geo-doc can assist in providing evidence of widespread systemic problems leading to a potentially more comprehensive and profound impact on progressive social change. The living documentary nature of the geo-doc also presents an incentive not to offend to avoid being added to the digital wall of shame that is permanently online and accessible to all, including current and potential employers, not to mention friends and family.

8.3 The Evolution of the Filmmaker

And where is the filmmaker in all this? As the form of the documentary changes, so too does the role of its makers. For the traditional filmed documentary, the director still exists, but no longer represented merely by their vision on the screen or their presence at film festivals. Today, many documentary filmmakers consider themselves activists and can be seen with their films at rallies, protests, and marches advancing the calls to action their films make. They are also becoming collaborators in a productive relationship with their intended audience of changemakers. In some cases, the filmmaker is given a seat at the table of policy creation (see Fig. 8.1) to serve as a surrogate for additional context to the visible evidence their documentaries provide to those charged with creating social change. In 2019 at COP25 in Madrid, indigenous youth filmmakers from the Inuit community of Tuktoyaktuk in Canada's Northwest Territories attended the conference presenting their film (*Happening to Us*) at meetings and side events. Hosted by the Youth Climate Report, their participatory role extended from behind and in front of the camera to the policy creation sessions of the United Nations Framework Convention on Climate Change.

It can also be argued that in addition to adopting a greater role for today's digital documentarian, there is also a vanishing of creative contribution and presence in the making of documentary projects for the filmmaker as the digital technology increasingly replaces them. Without a script and framed shots and with audiences entering the "film" rather than a cinema to simply view one, the roles of the director, writer, and editor—traditionally the principal contributors to the filmmaking process—are fading into the background along with their credits as they are now being replaced by "designers," "collaborators," and "artists." These are the new filmmakers of the remediated documentary as it continues to take shape in form and evolve as an influential communications tool of social change.

Since its origins as non-fiction film, the documentary has taken on many forms due to new theoretical approaches and technological advances. Many of these changes were made with the goal of enhancing the documentary film's ability to effect social change. This book has explored the documentary's ability to act as an instrument of social change through several lenses such as historical, theoretical, and digital, and has introduced more refined approaches such as multilinear formats, semiotic storytelling techniques, and geomedia platforms. Combining these into a sub-genre of the documentary—the geo-doc—provides a remediated form with the potential to create positive social change more effectively than its predecessors. No doubt, this evolution of its form and technology will continue as filmmakers, audiences, and changemakers use this influential communications tool in ways not yet imagined to assist their efforts in improving the societies, communities, and the world in which we live.

Bibliography

"About UNVR." *United Nations Virtual Reality*, January, 2015. Web. Accessed May 18, 2018. http://unvr.sdgactioncampaign.org/home/about.

Brucato, Ben. "Policing Made Visible: Mobile Technologies and the Importance of Point of View." *Surveillance & Society* 13, no. 3/4 (2015). Print.

Jenkins, Henry. "Charting Documentary's Futures: An Interview with MIT's William Uricchio (Part Four)." *Confessions of an Aca-Fan*. Henryjenkins.org: February 9, 2016. Web. Accessed May 20, 2017. http://henryjenkins.org/2016/02/charting-documentarys-futures-an-interview-with-mits-william-uricchio-part-four.html.

Tarnogol, Fernando. "What Is the Optimal 360 Video Length for Social Media?" *Medium* (March 13, 2016). Web. Accessed May 18, 2018. https://medium.com/@ftarnogol/what-is-the-optimal-360-video-length-for-social-media-d18e36b85d52.

Virtual Reality Projects

Antarctica in Decline, Director, Mark Terry. York University and United Nations Climate Change, 2017.

Building Brains, Building Futures, No Director Credited. UNICEF and Happy Finish, 2017.

Clouds Over Sidra, Directors, Gabo Arora and Barry Pousman. United Nations Sustainable Development Goals Action Campaign, Within, UNICEF Jordan, 2015.

Life in the Time of Refuge, Director, David Gough. United Nations High Commissioner for Refugees, Nokia, A Mad Production, 2017.

Afterword

Creating a geo-doc first requires a working knowledge of GIS software. There are many subscription-based and open-source programs currently available online. Many of them come with tutorials to train you in using their software. Some provide their software through a download feature, while others are accessible online. To get you started, I have compiled a list (without endorsement and in alphabetical order) of some of the more popular subscription-based and open-source programs currently available online along with their accompanying URLs:

Subscription-Based GIS

1. ArcGIS: https://www.arcgis.com
2. DMTI Spatial: https://www.dmtispatial.com
3. eSpatial: https://www.espatial.com
4. Geopointe: https://www.geopointe.com
5. Global Mapper: https://www.bluemarblegeo.com/products/global-mapper.php
6. GRASS GIS: https://grass.osgeo.org
7. Mapbox: https://www.mapbox.com
8. MapInfo Pro: https://www.pitneybowes.com/us/location-intelligence/geographic-information-systems/mapinfo-pro.html
9. MapMe: https://mapme.com

M. Terry, *The Geo-Doc*, Palgrave Studies in Media and Environmental Communication,
https://doi.org/10.1007/978-3-030-32508-4

10. MapRite: https://www.envitia.com/technologies/products/maprite
11. Maptitude: https://www.caliper.com/maptovu.htm
12. SAGA GIS: http://www.saga-gis.org
13. SuperGIS: https://www.supergeotek.com
14. Surfer: https://www.goldensoftware.com/products/surfer
15. TatukGIS: https://www.tatukgis.com
16. ZeeMaps: https://www.zeemaps.com

Open-Source GIS

1. Diva-GIS: http://www.diva-gis.org
2. GeoDa: http://geodacenter.github.io
3. Google My Maps: https://www.google.ca/maps/about/mymaps[1]
4. gvSIG: http://www.gvsig.com
5. GRASS GIS (Geographic Resources Analysis Support System): https://grass.osgeo.org
6. ILWIS (Integrated Land and Water Information Management): https://52north.org/software/software-projects/ilwis
7. MapWindow: https://www.mapwindow.org
8. OpenJump GIS: http://www.openjump.org
9. OrbisGIS: http://orbisgis.org
10. QGIS: https://qgis.org
11. uDig: http://udig.refractions.net
12. Whitebox GAT (Geospatial Analysis Toolbox): https://jblindsay.github.io/ghrg/Whitebox/index.html

The Youth Climate Report GIS Project was first created using Google Fusion Tables, a beta program Google decided to discontinue as of December 3, 2019. The project migrated to Google *My Maps*, a similar yet more robust and user-friendly version. Should you wish to test this software with a geo-doc project of your own, follow the step-by-step instructions divided into four phases of production provided below:

Phase I

1. Create a Google account.
2. Click "Create an account."

[1] Note: You must first have a Google account to access Google *My Maps*.

3. The signup form will appear.
4. Review Google's Terms of Service and Privacy Policy, click the checkbox, then click "Next Step."
5. The "Create Your Profile" page will appear.
6. Your account will be created, and the Google welcome page will appear.

Phase II

1. Open a new Excel Spreadsheet.
2. Create columns: for example, Title, Name, Description, Location, Video, Link(s), Year.[2]
3. For Location, provide the name of the city and province, state or country: for example, Toronto, ON; New York, NY; Paris, France—OR—search Google for the Latitude and Longitude coordinates (e.g. "Boston long and lat"). Copy and paste these numbers—*without* directions (N, S, E, W)—into the Locations column for more precise pin placement.
4. For Video, provide the YouTube URL from the video's "Share," *not* the webpage URL.
5. Save your Excel Spreadsheet as a "CSV" file (CSV stands for "comma-separated values").

Phase III

1. Sign-in to your Google Drive account.
2. Go to https://google.com/mymaps.
3. Select the red button labelled: "Create a new map."
4. Select: "Untitled map" and give your project a name.
5. Click on "Base map" to change the style of your project's map.
6. Click on "Import."
7. Select your CSV Excel Spreadsheet.
8. New Dialogue Box: "Choose columns to select your placemarks"—Select the box for "Location."
9. Click "Continue."
10. New Dialogue Box: "Choose a column to title your markers"—Select "Name."

[2]Other columns may be added to enhance your project, but these are recommended as essential to the basic structure of the geo-doc.

11. Click "Finish."
12. Your map is now created with pins indicating the place of your Locations and the related data fields.
13. To add Video, copy the URL from each pin's dialogue box.
14. Select the "camera" icon  .
15. New Dialogue Box: "Choose an image or video."
16. Select pull-down menu: "YouTube URL."
17. Select "YouTube URL."
18. Paste your YouTube URL in the box titled "Paste YouTube URL here."
19. Click "Select" button.

Phase IV

1. To play video from a pin, click the video thumbnail.
2. Click the word "YouTube."
3. Enlarge video by selecting frame icon  .

Once your geo-doc project has been created, you will be provided with a link and an embed code with which to share. The link can be dropped into any social media platform or email. The embed code is used to embed your map project directly into your website. It can also be provided to webmasters of partnering websites.

If you plan to use open-source software for your geo-doc, you may want to share the login information so others may make direct contributions to your map. Should you wish to control the content uploaded, provide an email address for others to direct content to you, then upload only those contributions that adhere to your project's format and goals.

GIS programs vary greatly in customizable features. For example, the shape and colour of your displayed pins can reflect informational features (in a climate change research model, blue can represent films and data related to oceans; green can represent forests; icons of film projectors can represent pins with video; icons of houses can represent urban centres, etc.). Many GIS programs allow you to upload your own icon to customize your map as well.

The column recommended for links can maximize your project's access to many of the digital affordances of the Internet. Here you can connect the user to websites featuring profiles on the interview subject featured in your film or websites of the organization or institution featured in your film. For the *Youth Climate Report GIS Project*, links to data and research papers published by the profiled interview subject are provided whenever possible so the user can access additional data to what was provided in the mini-documentary.

There is no limit as to what can be provided within the pins of a geo-doc project. The more that is made available, the easier it is for the change-maker to have a fuller understanding of the issue and write a policy that yields a progressive social change—often the goal of the activist documentary.

Index[1]

[1] Note: Page numbers followed by 'n' refer to notes.

M. Terry, *The Geo-Doc*, Palgrave Studies in Media and Environmental Communication,
https://doi.org/10.1007/978-3-030-32508-4

GPSR Compliance
The European Union's (EU) General Product Safety Regulation (GPSR) is a set of rules that requires consumer products to be safe and our obligations to ensure this.

If you have any concerns about our products, you can contact us on

ProductSafety@springernature.com

In case Publisher is established outside the EU, the EU authorized representative is:

Springer Nature Customer Service Center GmbH
Europaplatz 3
69115 Heidelberg, Germany

www.ingramcontent.com/pod-product-compliance
Ingram Content Group UK Ltd.
Pitfield, Milton Keynes, MK11 3LW, UK
UKHW022001190726
13853UKWH00004B/1666

* 9 7 8 3 0 3 0 3 2 5 1 0 7 *